A

STUDY GUIDE

FOR

DIALYSIS TECHNOLOGISTS

EDITED BY:

PHILIP M. VARUGHESE, BS, CHT, CCNT

JOAN ARSLANIAN, MS, MPA, MSN, RN, CS, FNP-BC, CNN, CHN, CPDN

JOHN A. SWEENY, BS, CHT

A STUDY GUIDE FOR DIALYSIS TECHNICIANS/TECHNOLOGISTS
FIFTH EDITION

EDITED BY:

PHILIP M. VARUGHESE, BS, CHT, CCNT
Facility Administrator
DaVita-Richmond Kidney Center
Staten Island, New York

JOAN ARSLANIAN, MS, MPA, MSN, RN, CS, FNP-BC, CNN, CHN, CPDN
Clinical Director/Nurse Practitioner
The New York Hospital Queens
Trude Weishaupt Memorial Dialysis Center
Fresh Meadows, New York

JOHN A. SWEENY, BS, CHT
Technical Training Manager
Baxter Healthcare Inc.
Largo, Florida 33773

ISBN: 978-0-9835088-1-6

Cover Illustration:
The illustration on the cover shows the principle of diffusion, the main driving force for the removal of small molecules such as urea and creatinine from the patient's blood during a hemodialysis treatment. The green square represents the dialysate with some of its constituents: water, bicarbonate, and sodium chloride. The yellow box represents the effluent from the dialyzer. In addition to the dialysate molecules, it also contains urea (blue squares), creatinine (blue circles), and a few mid-size molecules (black diamonds). Inside the hollow fiber membrane are large circles that represent red blood cells. These are too large to cross the membrane and hence, do not appear in the effluent solution. The diffusion performance in a modern hollow fiber dialyzer is so efficient that for every four urea molecules entering the top of the dialyzer, only one make it through the dialyzer to return to the patient in spite of the fact that the blood only spends about 15 to 20 seconds in the dialyzer.

FOREWORD

The career of the hemodialysis technician has grown in response to the needs of patients seeking hemodialysis as a modality of treatment for end stage renal disease. A new edition of *A Study Guide for Dialysis Technologists* will help to meet the educational needs of the hemodialysis professional.

The need for this book is clear because of the paucity of such literature in the nephrology community. This comprehensive guide to performance of procedures and day-to-day care of the hemodialysis patient is on the cutting edge of technological and legislative changes as the dialysis professional strives to keep current in a changing industry.

This latest edition will appeal to every member of the nephrology health care team and will be an invaluable tool for revising existing documents, teaching dialysis team members and referencing information for educators. Although the topics are complex the authors have simplified it to appeal to entry level practitioners. Its diagrams, charts, abbreviations and conversion tables are designed for everyday use, as those in the dialysis arena are faced with challenging clinical situations.

As president of the Board of Nephrology Examiners Nursing and Technology that certifies nurses and technicians, I think that this book is pitched at a level that should assist candidates as they prepare for the exams. This fifth edition will provide a current perspective on diverse topics and interests that nephrology professionals will find invaluable.

RJ Picciano, BA, CHT, OCDT
BONENT President

PREFACE

With the birth of this 5th edition, I am thrilled that under the Interpretive Guidance: New Conditions for Coverage for ESRD facilities, all dialysis technicians need national certification to practice in the US as of April 15, 2010. The field of dialysis technology has changed by leaps and bounds since the publication of the 3rd edition in 2003, and finally, the role of dialysis technician in patient care is beginning to get the recognition it so long deserved.

This 5th edition has been revised to reflect the changes in the dialysis field over the last six years and to meet the new requirements set forth by CMS for conditions for coverage. This edition is meant not to only serve as a review tool, but also as a great information resource for new as well as seasoned technicians. My hope is that this review guide will provide a wealth of information to technicians no matter what their level of experience or years of practice.

The illustration on the front cover was created by Clement Leduc, a hemodialysis systems specialist for Baxter Healthcare Inc. in Ottawa, Ontario. It is used to educate technicians new to the field of dialysis about the process of diffusion of molecules utilizing a hollow fiber dialyzer during a hemodialysis treatment.

This 5th edition was published and made possible through the continued support of the National Association of Nephrology Technicians/Technologists (NANT), an organization for which I am honored to have served as President. Back then, it seemed like a dream and a far-reaching goal that dialysis technicians would attain the level of professionalism it has reached today. My sincere thanks to the NANT Board of Directors and especially Francine Rickenbach, NANT Executive Director, for supporting our efforts to educate and empower dialysis technicians through the years.

I would also like to thank Joan Arslanian and John Sweeny for their countless hours spent reviewing and contributing content to this edition. I am also indebted to RJ Picciano, BONENT President for writing the foreword for this edition and for all her support. My heartfelt thanks also goes to Lenore Arato who spent many hours typing and editing this guide for me.

I saved the best for last. My wife, Alamma, has patiently supported all my professional endeavors over the years, and this was no different. As I spent hours that never seemed to end writing this edition, she stood by me and encouraged me all the way. Words aren't enough to express the depth of my gratitude.

Philip Varughese
Facility Administrator
Richmond Kidney Center
Staten Island, New York
June 2012

TABLE OF CONTENTS

Table of Contents

Tables & Illustrations

Section I
RENAL ANATOMY & PHYSIOLOGY

1. A major body fluid in the intravascular compartment is:
   A) Plasma
   B) Serum
   C) Water
   D) Saline

2. The major vessel that supplies blood to the kidney is the:
   A) Aorta
   B) Inferior vena cava
   C) Renal vein
   D) Renal artery

3. The volume of plasma filtered from the glomerular capillaries into Bowman's capsule each minute is the:
   A) Tubular reabsorption
   B) Glomerular Filtration rate
   C) Tubular secretion
   D) Clearance

4. What percent of the cardiac output each minute is received by the kidneys?
   A) 25%
   B) 15%
   C) 99%
   D) 50%

5. The endocrine gland located on top of each kidney is the:
   A) Pituitary
   B) Lymphatic
   C) Adrenal
   D) Medulla

6. The functional unit of the kidney is the:
   A) Adrenal gland
   B) Cortex
   C) Medulla
   D) Nephron

7. The main function of the glomerulus is:
   A) Secretion
   B) Filtration
   C) Reabsorption
   D) Excretion

8. Three steps in the formation of urine are:
A) Glomerular filtration, diffusion, ultrafiltration
B) Filtration, reabsorption, ultrafiltration
C) Filtration, reabsorption, secretion
D) Clearance, glomerular filtration, ultrafiltration

9. The normal adult glomerular filtration rate is approximately:
A) 99 mL/hr
B) 2000 mL/day
C) 80 mL/hour
D) 125 mL/min

10. Which of the following substances is NOT filtered in Bowman's space?
A) Chloride
B) Sodium
C) Water
D) Proteins

11. One of the effects of the renin-angiotensin system is to:
A) Stimulate the release of aldosterone
B) Decrease serum sodium
C) Decrease extracellular fluid volume
D) Dilate the arteries

12. Glomerular filtration begins as blood enters the glomerulus under high pressure called:
A) Capillary colloidal osmotic pressure
B) Glomerular capillary hydrostatic pressure
C) Tubular hydrostatic pressure
D) Ureteral pressure

13. Which of the following is NOT a function of parathyroid hormone?
A) Decreases reabsorption of calcium from intestine
B) Stimulates the activation of Vitamin D
C) Increases reabsorption of calcium from the bone
D) Decreases reabsorption of phosphate by tubules

14. The part of the nephron where most of the glomerular filtrate is reabsorbed is:
A) Bowman's capsule
B) Distal tubule
C) Loop of Henle
D) Proximal convoluted tubule

15. The two components of the nephron are:
A) Vascular, tubular
B) Cortex, medulla
C) Vascular, collecting system
D) Adrenal, tubular

16. All of the following are excretory functions of the kidney EXCEPT:
A) Regulation of electrolytes
B) Maintenance of fluid balance
C) Elimination of metabolic waste
D) Release of aldosterone

17. Diffusion is the movement of:
A) Solute from an area of high concentration to an area of low concentration until equilibrium is reached
B) Solute from an area of low concentration to an area of high concentration until equilibrium is reached
C) Solvent from an area of high concentration to an area of low concentration until equilibrium is reached
D) Solvent from an area of high concentration to an area of low concentration until equilibrium is reached

18. ADH (antidiuretic hormone) regulates the:
A) Chloride content of the body
B) Potassium content of the body
C) Water content of the body
D) Magnesium content of the body

19. The hormone aldosterone stimulates:
A) Reabsorption of sodium and excretion of potassium
B) Reabsorption of sodium and potassium
C) Excretion of sodium and potassium
D) Reabsorption of potassium and excretion of sodium

20. The hydrogen ion concentration determines the acid-base balance of the body. Hydrogen ion concentration is affected by the ratio of:
A) Carbonic acid to bicarbonate
B) Acid hemoglobin to basic hemoglobin
C) Acid phosphate to alkaline phosphate
D) Red corpuscles to white corpuscles

21. The kidneys' role in calcium/phosphate balance is the:
    A) Activation of Vitamin D
    B) Excretion of phosphorus
    C) Reabsorption/excretion of calcium
    D) All of the above

22. Calcium absorption occurs in the:
    A) Stomach with phosphorus
    B) Small intestine with activated Vitamin D
    C) Large intestine with Vitamin D
    D) None of the above

23. Parathyroid hormone (PTH) regulates calcium/phosphorus levels by:
    A) Stimulating the kidney to excrete phosphorus
    B) Stimulating the activation of Vitamin D
    C) Increasing bone resorption
    D) All of the above

24. Clearance can be defined as:
    A) A substance that is filtered at the glomerulus and secreted in the Loop of Henle
    B) Volume of plasma that is cleared of a given substance per unit of time
    C) A substance that is filtered, secreted, and reabsorbed
    D) Volume of water that is cleared per unit of time

25. A factor(s) that affect the level of BUN is (are):
    A) Protein intake
    B) Steroids
    C) Infection
    D) All of the above

Use the following diagram to answer questions 26-31.

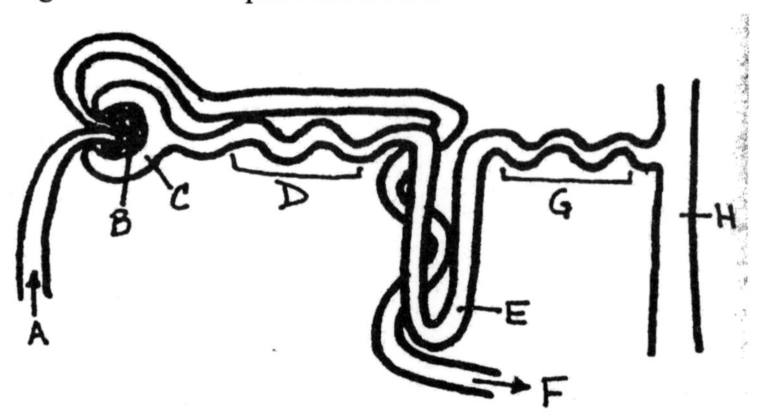

Identify the following parts of the nephron using the diagram on the previous page.

26. ____ Glomerulus
27. ____ Distal convoluted tubule
28. ____ Collecting Duct
29. ____ Loop of Henle
30. ____ Bowman's capsule
31. ____ Proximal convoluted tubule

Match the acid-base imbalance with its chemical manifestation.

32. ____ Respiratory acidosis
33. ____ Metabolic acidosis
34. ____ Respiratory alkalosis
35. ____ Metabolic alkalosis

A) Increase in P_{CO2}
B) Decrease in HCO_3
C) Decrease in P_{CO2}
D) Increase in HCO_3

36. Normally, the lungs work to keep the P_{CO2} at approximately 20 mmHg and the kidneys keep the HCO_3 level between 24-26 mEq/L.
A) True
B) False

37. When one HCO_3 ion is lost from the kidney, three H ions stay behind
A) True
B) False

38. Bicarbonate is reclaimed by the proximal tubule while hydrogen ions are excreted by collecting tubules.
A) True
B) False

39. The substance secreted by the kidney, which affects the blood pressure, is:
A) Potassium
B) PTH
C) Aldosterone
D) Renin

40. Sodium is primarily an __________ ion.
A) Intracellular
B) Extracellular

41. Potassium is primarily an _________ ion.
A) Intracellular
B) Extracellular

42. Normal carbon dioxide tension (P_{CO2}) found in arterial blood is:
A) 20-30 mmHg
B) 22-26 mmHg
C) 35-45 mmHg
D) 45-55 mmHg

43. Normal oxygen tension (P_{O2}) found in arterial blood is:
A) 20-30 mmHg
B) 40-50 mmHg
C) 75-85 mmHg
D) 75-100 mmHg

44. Seventy percent (70%) of the total body water is contained in organs.
A) True
B) False

45. Urea nitrogen is the by-product of:
A) Protein metabolism
B) Phosphorus absorption
C) Calcium metabolism
D) Muscle metabolism

46. Under normal conditions, the plasma calcium/phosphate product is maintained at about 40mg/dL. The patient with renal failure loses the ability to maintain a normal product because of:
A) Altered renal excretion of phosphate
B) Increased GI absorption of calcium
C) Altered deposition of phosphate in bone
D) Increased thyrocalcitonin production

47. The changes related to uremia develop from:
A) The kidney's inability to maintain the basement membrane
B) The effects of hemolytic anemia
C) The kidney's inability to maintain the internal environment
D) The effects of altered antidiuretic hormone function

48. Decreased glomerular filtration rate of the kidney may be due to all of the following <u>EXCEPT</u>:
A) Fluid overload
B) Dehydration
C) Shock
D) Cardiac decompensation

49. What is the normal range for glomerular filtration rate?
A) 120-130 mL/min
B) 1-5 mL/min
C) 500-600 mL/min
D) 180-200 mL/min

50. Homeostasis can best be defined as:
A) The steady state of the internal environment
B) A holistic approach to medicine
C) A balance between nature and man
D) The state of physical and psychological well-being

51. The main function(s) of the nephron tubules is:
A) Reabsorption
B) Secretion
C) Filtration
D) A & B

52. According to the National Kidney Foundation, anyone with a glomerular filtration rate less than _70_ mL/min/1.73m^2 for more than three months is defined as having chronic kidney disease.
A) 100
B) 90
C) 80
D) 70

53. The major components of a nephron unit are:
A) Papilla, hilum, peritubular capillary network
B) Cortex, capsule, juxtaglomerular complex
C) Glomerulus, peritubular capillary network, tubule
D) Medulla, glomerulus, afferent arteriole

54. Under normal conditions, glomerular filtrate does not contain which of the following substances?
A) Amino acids
B) Red blood cells
C) Glucose
D) Creatinine

55. The presence of _____ in the urine is often indicative of glomerular capillary damage.
A) Excess sodium
B) Glucose
C) Protein (albumin)
D) Casts

56. The normal kidney plays a role in red blood cell production by:
A) Stimulating white blood cell maturation in bone marrow
B) Serving as a reservoir for new red blood cells
C) Secreting a hormone called erythropoietin
D) Eliminating macrocytes

57. Toxic products that accumulate in the body as a result of decreased kidney function are:
A) Urea, creatinine, bile, and salt
B) Urea, creatinine, potassium, and phosphorus
C) Potassium, urea, amylase, and water
D) Sodium, urea, potassium, and alkali

58. In the presence of active Vitamin D, this substance can be properly absorbed and deposited in the skeletal system:
A) Phosphorus
B) PTH
C) Calcium
D) Sodium

59. The hormone erythropoietin causes:
A) Loss of kidney function
B) Blood pressure control
C) Regulation of body fluids
D) Bone marrow to produce red blood cells

Section II
CAUSES OF RENAL FAILURE

1. Glomerulonephritis can best be described as:
   A) An infectious process that involves Bowman's capsule and peritubular capillaries
   B) An inflammatory process that primarily affects the glomerular capillaries
   C) An obstructive process of the renal arterioles
   D) A congenital process that causes sclerosis of the tubules

2. As renal function deteriorates, laboratory data that reflects this is:
   A) an increase in BUN, increase in hemoglobin, decrease in creatinine, increase in creatinine clearance
   B) an increase in BUN, a decrease in hemoglobin, increase in serum creatinine, decrease in creatinine clearance
   C) a decrease in BUN, decrease in hemoglobin, increase in creatinine, decrease in creatinine clearance
   D) a decrease in BUN, increase in hemoglobin, increase in creatinine, increase in creatinine clearance

3. The most common cause of chronic renal failure in the U.S. is:
   A) Diabetes Mellitus
   B) Hypertension
   C) Polycystic kidney disease
   D) Glomerulonephritis

4. A kidney disease that usually presents with abdominal fullness, hematuria, polyuria but continued erythropoietin production is:
   A) Falconi syndrome
   B) Wilm's tumor
   C) Polycystic kidney disease
   D) Glomerulonephritis

5. A systemic disease characterized by an accumulation of collagen in connective tissue is:
   A) Glomerulonephritis
   B) Scleroderma
   C) Amyloidosis
   D) Retroperitoneal fibrosis

6. A chronic systemic inflammatory disorder of connective tissue especially in the arterial vasculature that results from the formation of auto antibodies is:
   A) Systemic lupus erythematosus
   B) Good Pastures syndrome
   C) Nephrotic syndrome
   D) Glomerulonephritis

7. A patient presents in the ER with hyperkalemia. Dialysis is not readily available. An interim treatment that may be used is:
A) Kayexelate
B) Glucose and insulin
C) IV sodium bicarbonate
D) All of the above

8. The accumulation of nitrogenous waste products in the bloodstream is called:
A) Uremia
B) Acute kidney failure
C) Chronic renal failure
D) Azotemia

9. Which of the following can cause renal failure?
I. Heroin
II. Tobramycin
III. Excessive intake of high potassium foods
A) I only
B) II only
C) I and II only
D) All of the above

10. Acute renal failure differs from chronic renal failure in that acute renal failure:
A) There is a low mortality rate
B) Usually requires dialysis therapy
C) Is monitored by hyperkalemia and fluid overload
D) Is sudden onset with reversible damage

11. All of the following are common causes of chronic renal failure <u>EXCEPT</u>:
A) Diabetes
B) Hypernatremia
C) Hypertension
D) Glomerulonephritis

12. Chronic kidney disease (CKD) Stage 5 is characterized by:
I. Residual renal function less than 15% of normal
II. Excretory, regulatory, and hormonal renal functions that are severely impaired
III. Inability to maintain homeostasis
IV. Markedly elevated BUN and creatinine
A) I and II only
B) II and III only
C) III and IV only
D) All of the above

13. An endocrine disease which affects the kidney is:
A) Lupus
B) Hypertension
C) Hypotension
D) Diabetes

14. Ms. T. has Goodpasture Syndrome, which is:
A) More common in women than in men
B) Characterized by the presence of autoimmune anti-basement membrane antibodies to glomerular & alveolar capillaries
C) Primarily a childhood disorder
D) All of the above

15. The National Kidney Foundation, through its Kidney Disease Outcomes Quality Initiative guidelines, stages kidney disease according to:
A) Severity of symptoms
B) Results from a 24-hour urine collection
C) Results of spot urine albumin-to-creatinine ratio
D) Estimated glomerular filtration rate (GFR)

16. Nephrotoxic substances include all of the following <u>EXCEPT</u>:
A) Excessive amounts of water
B) Radiopaque contrast media
C) Antibiotics
D) Anesthesia

17. Which of the following populations are at risk for renal disease?
A) Elderly
B) Trauma victims
C) Diabetes Mellitus
D) All of the above

Section III
CLINICAL MANIFESTATIONS OF RENAL FAILURE

1. Management of pericarditis might include all of the following EXCEPT:
   A) Aggressive blood pressure control
   B) Controlled heparinization during dialysis
   C) Increased frequency of dialysis
   D) Use of anti-inflammatory drugs

2. The presence of a pericardial friction rub may indicate:
   A) Cardiac overload
   B) Uremic pericarditis
   C) Pericardial tamponade
   D) Mild pericardial effusion

3. Mechanisms precipitating hypocalcemia in the patient with chronic renal failure include:
   A) Hyperphosphatemia
   B) Hypophosphatemia
   C) Decreased parathyroid hormone secretion
   D) Increased Vitamin D secretion

4. Speech disturbance is usually the first and foremost clinical feature of:
   A) Dialysis encephalopathy/dementia
   B) Lead poisoning
   C) Metabolic acidosis
   D) Dialysis disequilibirium

5. A patient's Ca^{++} X PO_4^{---} product is above normal, the practitioner would pay particular attention to signs of which of the following:
   A) Neuromuscular disease
   B) Connective tissue disease
   C) Myocardial conduction disease
   D) Bone disease

6. Anemia is due to:
   - I. Inadequate red blood cell production
   - II. Blood loss
   - III. Abnormal destruction of red blood cells
   - IV. A combination of abnormalities

   A) I, II, and IV only
   B) II, III, and IV only
   C) I, III, and IV only
   D) All of the above

7. Symptoms of anemia can include:

    I. Angina
    II. Dyspnea
    III. Loss of appetite
    IV. Weakness

    A) I and II only
    B) I and III only
    C) II and IV only
    D) All of the above

8. The anemia of uremia, without iron deficiency, is classified as being:
    A) Normocytic
    B) Microcytic
    C) Macrocytic
    D) None of the above

9. A clinical manifestation of a patient in acute metabolic acidosis is:
    A) Hyperventilation
    B) Malignant hypertension
    C) No change in clinical status
    D) Hypovolemia

10. EKG changes indicative of hyperkalemia would include:
    A) Prolonged PR intervals
    B) Wide QRS complexes
    C) Tall tented T-waves
    D) All of the above

11. The major cause of anemia in chronic renal failure is:
    A) Frequent blood sampling
    B) Gastrointestinal bleeding
    C) Decreased RBC production
    D) Poor dietary iron intake

12. A complaint of severe tingling and numbness in a patient's toes and fingers is most likely:
    A) Guillain Barre
    B) Shingles
    C) Peripheral neuropathy
    D) Arteriosclerosis

13. Amyloidosis is caused by:
   A) A streptococcal organism
   B) Deposits of glycoprotein
   C) An accumulation of collagen
   D) None of the above

14. A pale, grayish bronze skin color often seen in a chronic renal failure patient is caused by:
   A) Inadequate dialysis
   B) Anemia due to lack of erythropoietin
   C) Retained pigments normally excreted
   D) Decreased oil gland activity

15. Dry, itchy skin in patients with chronic renal failure is caused by:
   A) Deposits of calcium phosphate crystals
   B) Decreased oil gland activity
   C) Irritable nerve endings
   D) All of the above

16. A white, powdery substance, which deposits on the skin surfaces with perspiration in uremic individuals, is called:
   A) Calcium phosphate crystals
   B) Uremic frost
   C) Ecchymosis
   D) Purpura

17. Hyperkalemia in chronic renal failure can be caused by:
   I. Dietary intake of high potassium containing foods
   II. Administration of blood transfusions
   III. Bleeding, hemolysis
   IV. Decreased urinary excretion

   A) I and III only
   B) I and IV only
   C) I, III, and IV only
   D) All of the above

18. The acid-base imbalance seen in a chronic renal failure patient is:
   A) Respiratory alkalosis
   B) Metabolic alkalosis
   C) Respiratory acidosis
   D) Metabolic acidosis

19. A cause(s) of pericarditis in renal failure patients is (are):
   A) Inadequate dialysis prescription
   B) Missed dialysis treatments
   C) Fluid overload
   D) All of the above

20. Symptom(s) manifested by patients with pericarditis is (are):
   A) Chest pain
   B) Friction rub
   C) Low grade fever
   D) All of the above

21. A complication of pericarditis which can occur while the patient is on hemodialysis is:
   A) Pneumonia
   B) Hypertension
   C) Hypothermia
   D) Cardiac tamponade

22. Signs and symptoms of the complications in Question #21 is (are):
   A) Change in mental status
   B) Hypotension
   C) Muffled heart sound
   D) All of the above

23. Treatment for a patient with pericarditis on hemodialysis includes:
   A) No heparin
   B) Increasing the intensity of dialysis
   C) Increasing fluid intake to increase circulating volume
   D) A and B

24. Chronic renal failure patients are prone to infections due to:
   I. Lack of erythropoietin
   II. Decreased life span of platelets
   III. Dry/itchy skin
   IV. Decreased life span and defective white blood cells

   A) II, III, and IV only
   B) I and IV only
   C) III and IV only
   D) All of the above

25. Possible cause(s) of constipation in a patient with chronic renal failure is (are):
A) Lack of vegetables and fruits in the diet
B) Fluid restriction
C) Phosphate binders
D) All of the above

26. Sexual impotence that is sometimes experienced by a patient in renal failure may be due to:
A) Low testosterone
B) Low oxygenation
C) Antihypertensive medications
D) All of the above

27. Neurological manifestations that may be experienced in the renal failure patient is (are):
A) An inability to concentrate
B) A decreased intellectual capacity
C) An increased memory
D) All of the above

28. Restless legs and burning feet experienced by patients with chronic renal failure is caused by:
A) Depression
B) Hypokalemia
C) Hypernatremia
D) Nerve end damage

29. Aluminum toxicity, metabolic acidosis, and hyperparathyroidism may contribute to which long-term effect in the patient with chronic renal failure?
A) Pericarditis
B) Respiratory insufficiency
C) Bone demineralization
D) Peripheral neuropathy

30. Diabetic nephropathy causes:
A) Sclerosing of the arterioles in the nephron
B) Damage to the tubular components of the nephron
C) Loss of concentrating ability of the distal tubule
D) Loss of GFR due to proximal tubule damage

31. All of the following may be seen in hyperparathyroidism <u>EXCEPT</u>:
A) Elevated calcium level
B) Elevated phosphorus level
C) Decreased PTH level
D) Elevated PTH level

32. The long-term effect of hyperlipidemia in chronic renal failure patients is:
A) Abnormal insulin production
B) Obesity
C) Respiratory insufficiency
D) Cardiovascular disease

33. Management of dry, itchy skin in the patient with chronic renal failure includes all of the following EXCEPT:
A) Good hygiene and lubrication of skin
B) Phosphate binders
C) Adequate heparinization
D) Adequate dialysis

34. Hyperkalemia can result in:
A) Bleeding
B) Bradycardia
C) Acidosis
D) Hypophosphatemia

35. Fluid and sodium overload, and abnormalities in renin production and excretion can cause:
A) Hyperkalemia
B) Acidosis
C) Hypertension
D) Bone disease

36. Neurological changes that occur with uremia include:
I. Anxiety
II. Depression
III. Restless leg movements
IV. Agitation
V. Insomnia
A) I, II & III
B) I, II & IV
C) II, III & IV
D) III, IV & V

37. Ms. Smith complains of severe pruritus. The practitioner suggests that she:
A) Use superfatted soaps and apply lotion liberally
B) Increase her high biologic value protein intake
C) Increase her intake of dairy products, such as milk & cheese
D) Decrease her dosage of phosphate binding medication to reduce her plasma calcium-phosphate product

38. Which complication(s) of CKD are also risk factors for development of cardiovascular disease?
A) Hypertension only
B) Hypertension & anemia only
C) Hypertension, anemia & disorders of bone metabolism only
D) Hypertension, anemia, disorders of bone metabolism & proteinuria

39. Inflammation of the pericardium resulting in an accumulation of fluid in the pericardial sac is called:
A) Peritonitis
B) Pleural effusion
C) Pericarditis
D) Pulmonary edema

40. Uremic patients tend to bleed, or develop spontaneous bruises primarily related to:
A) Anemia
B) Platelet dysfunction
C) Hyperkalemia
D) Under heparinization

41. Edema may be defined as:
A) An increase in the amount of fluid in the extracellular tissues
B) An increase in the amount of extracellular tissue
C) A decrease in the amount of extracellular tissues
D) A decrease in the amount of intracellular fluid

42. Symptoms of anemia are likely to manifest when the Hct/Hgb level declines to approximately what percentage of the mean level for the normal healthy respective age and gender group?
A) 80
B) 60
C) 50
D) 30

43. Erythropoetin helps regulate:
A) Sodium reabsorption
B) Circulating hemoglobin
C) Blood pressure
D) Water reabsorption

44. The anemia of renal failure is an ongoing complication for many patients with CKD. Which of the following best describes a practitioner's actions <u>during</u> dialysis toward minimizing the anemia?
A) Minimize blood loss through proper technique
B) Give vitamin supplements during dialysis
C) Perform hemoglobin and hematocrit studies with each treatment
D) Request a dietary consult

45. Erythropoietin:
A) Is a hormone secreted by the kidney which stimulates red blood cell production in bone marrow
B) Is a hormone secreted by bone marrow which limits destruction of red blood cells in the kidney
C) Is an enzyme released by activated (killer) T-lymphocytes which minimizes hemolysis
D) Is an enzyme which converts angiotensin I to angiotensin II

46. The primary reason for anemia in chronic renal failure is:
A) Decreased erythropoietin production
B) Platelet dysfunction
C) Increased bleeding tendencies
D) Loss of blood during hemodialysis

47. A proper balance of this chemical is essential for normal muscle function; an excess amount can cause the heart to stop beating:
A) Glucose
B) Creatinine
C) Phosphorus
D) Potassium

48. The patient in chronic renal failure often complains of severe pruritis. The nurse would instruct the patient to:
A) Use superfatted soaps & lotions
B) Increase dietary protein intake
C) Increase dietary intake of foods high in phosphate
D) Use glycerine to prevent itching

49. Parathormone normally helps to maintain a normal serum calcium level. This hormone from the parathyroid gland acts on:
A) Bone osteoblasts to build new bone
B) Renal tubules to decrease phosphate reabsorption
C) Bone osteoblasts to increase bone resorption
D) Gastrointestinal cells to increase calcium content in feces

50. In chronic renal failure, multiple complex derangements occur in calcium metabolism. As a result, patients experience:
A) Secondary hyperparathyroidism
B) Metastatic calcification
C) Osteomalacia and osteitis fibrosis
D) All of the above

51. Chronic renal failure affects the patient's neurologic function. Specific problems which must be evaluated include:
A) shortened attention span
B) decreased level of consciousness
C) confusion
D) all of the above

Section IV
PSYCHOSOCIAL

1. A stress factor(s) that may influence the adaptation of the CKD patient to their disease is (are):
   A) Financial
   B) Body image changes
   C) Effects of the disease
   D) All of the above

2. In the stages of adaptation as described by Reisman & Levy, during the honeymoon phase, the patient typically presents as:
   A) Discouraged and disenchanted
   B) Angry and hostile
   C) Hopeful and confident
   D) All of the above

3. Adaptation can be described as:
   A) Using palliative behaviors
   B) Non-compliance
   C) Goal setting
   D) Minimizing problems

4. All of the following defense mechanisms can be unhealthy if used on a long-term basis, <u>EXCEPT</u>:
   A) Rationalization
   B) Displacement
   C) Denial
   D) Problem solving

5. A behavior which would indicate adaptation to dialysis might be:
   A) Maneuvering dialysis therapy around work life
   B) Learning about medications
   C) Missing dialysis treatments
   D) Rationalization

6. The patient with chronic renal disease is told he must go on hemodialysis because his disease cannot be cured. What coping mechanism is the patient most likely to use initially?
   A) Regression
   B) Reaction formation
   C) Denial
   D) Projection

7. Grief is an expected phase in the process of adaptation to chronic renal failure. This is primarily due to:
   A) Anxiety about the unexpected
   B) Crisis in their life
   C) Fear of the unknown
   D) Loss of prior roles/relationships

8. Palliative coping behaviors often seen in the patient may be:
   A) Avoidance
   B) Taking medications
   C) Asking questions
   D) All of the above

9. The following is considered to be a positive factor in maintaining an individual's health?
   A) Control
   B) Commitment
   C) The ability to meet challenges
   D) All of the above

10. Social affiliation and expression of needs has been found to have a positive effect on illness.
   A) True
   B) False

11. The following "coping" style(s) assists in limiting illness.
   A) Approach
   B) Avoidance
   C) Denial
   D) All of the above

12. The following psycho-spiritual factor(s) aids in improving healing.
   A) Positive emotions
   B) Desire and will to love
   C) Changing behavior and habits
   D) All of the above

13. An individual's denial of negative emotion will foster a disease state.
   A) True
   B) False

14. Anger is an emotion that may surface after an individual is diagnosed with a disease state.
A) True
B) False

15. The following psychosocial factor(s) was/were found to be significantly beneficial to the treatment of illness.
A) Being hopeful and optimistic
B) Confidence in the treating physician
C) Emotional support from family and friends
D) All of the above

16. When faced with an illness and ongoing treatment, which of the following could become a stressor(s)?
A) Financial status
B) Self image
C) Family relationships
D) All of the above

17. A generalized state of discomfort, uncertainty, or fear stemming from an actual or potential threat to physiological and psychological well-being is:
A) Denial
B) Anxiety
C) Repression
D) Rationalization

18. The extent to which a person's behavior coincides with health advice is called:
A) Non-compliance
B) Denial
C) Compliance
D) Anxiety

19. Common psychological reactions & coping mechanisms of patients on maintenance dialysis include:
I. Anxiety
II. Depression
III. Hostility
IV. Anger
V. Denial
A) I, II, III, IV
B) I, II, IV, V
C) II, III, IV, V
D) I, II, III, IV, V

Section V
HEMODIALYSIS

1. Solutions that have a higher osmolarity than body fluids are called:
   A) Hypotonic
   B) Hypertonic
   C) Isotonic
   D) Hypo-osmole

2. Blood that is exposed to a hypertonic solution may cause:
   A) Hemolysis
   B) Crenation
   C) Hypertension
   D) Hyperthermia

3. Hemostasis depends on:
   I. The ability of normal blood vessels to contract and retract when injured
   II. The ability of platelets to form plugs in the injured blood vessels
   III. The presence of plasma coagulation factors
   IV. The stability of the blood clot
   A) I and II
   B) III and IV
   C) I, II, and III
   D) I, II, III, and IV

4. A non-sterile aqueous solution that is similar to normal levels of electrolytes found in the extracellular fluid is:
   A) Dialysate solution
   B) Plasma
   C) Normal saline
   D) Hypertonic saline

5. Dry weight is the term used to describe the status of a patient with no fluid excess in the:
   A) Interstitial and vascular compartment
   B) Intracellular compartment
   C) Intracellular and interstitial compartment
   D) Vascular compartment

6. The principle underlying waste product removal process in hemodialysis is:
   A) Osmotic pressure
   B) Hydrostatic pressure
   C) Filtration
   D) Diffusion

7. The movement of solute particles from an area of higher chemical concentration to an area of lower chemical concentration until equilibrium is reached is called:
   A) Diffusion
   B) Osmosis
   C) Ultrafiltration
   D) Hydrostatic pressure

8. Blood and dialyzing fluid flowing in opposite directions is called:
   A) Counter-current flow
   B) Cross-current flow
   C) Con-current flow
   D) Capillary flow

9. Factors which influence dialyzer clearance include:
   A) Blood flow rate
   B) Dialysate flow rate
   C) Membrane permeability
   D) All of the above

10. Which of the following is necessary for high flux dialysis?
   A) Sodium modeling
   B) Sterile dialysate
   C) Ultrafiltration control
   D) Kinetic modeling capabilities

11. In a roller-type pump, the pressure of the rollers against the tubing is called:
   A) Constriction
   B) Compression
   C) Impaction
   D) Occlusion

12. An air/foam detector is used to detect the presence of air in the:
   A) Heparin line
   B) Dialysate line
   C) Venous blood line
   D) Concentrate line

13. Pre-pump arterial pressure should be measured, particularly with high blood flow rates because:
   A) Pressure may be high (>100 mmHg)
   B) High ultrafiltration rates may damage red blood cells
   C) Hemolysis may occur with very low (more negative) undetected negative pressures
   D) Blood flow will improve

14. Pre-pump arterial pressure in a patient with a fistula should:
 A) Be below zero (negative)
 B) Be above zero (positive)
 C) Have the upper alarm limit set below zero
 D) A and C

15. Sequential ultrafiltration is BEST described as:
 A) Increased fluid removal using 2 dialyzers in series during one dialysis
 B) Increased fluid removal by alternating high and low TMP every hour during one dialysis
 C) Increased fluid removal in two successive dialysis
 D) Fluid removal without diffusion dialysis

16. The purpose of priming the extracorporeal circuit with saline during set up is:
 A) To remove sterilant, manufacturing residues or plasticizers before initiating treatment
 B) To be certain fibers are thoroughly wetted before initiating dialysis
 C) To assess the ultrafiltration capacity of the dialyzers and check for leaks
 D) To maintain fiber wetness until the treatment starts

17. Patient safety depends on properly proportioned dialysate. This is measured by total conductivity. Total conductivity refers to:
 A) The overall conductance of all ions in solution
 B) The conductance of sodium, calcium, potassium, and bicarbonate in solution
 C) The amount of electricity generated by free solids in a solution
 D) The electrical charge of chloride ions in dialysate

18. Hemolysis can occur during dialysis if:
 A) The dialysate concentration is too low, causing the cells to swell and burst as they take on water in an attempt to equalize their concentration of electrolytes with that in solution
 B) The temperature of dialysate is too low, causing the cells to contract in an attempt to conserve energy
 C) The dialysate concentration is too high, causing the cells to swell and burst as they take on water in an attempt to equalize their concentration of electrolytes with that in solution
 D) The potassium concentration of the dialysate is too low. This situation is especially hazardous to digitalized patients

19. A patient gains 2.5kgs since his last hemodialysis treatment. How much extra fluid does he have?
 A) 1000cc
 B) 1500cc
 C) 2000cc
 D) 2500cc

20. When fluid replacement is necessary during dialysis, which one of the following solutions is usually used?
A) 0.45% saline
B) 0.9% saline
C) 5% Dextrose in water
D) 5% dextrose in normal saline

21. Dry weight is defined as the weight gain at which the patient is free of swelling, other signs of fluid excess, and:
A) The temperature is normal when standing
B) The blood pressure, when standing, is normal for the patient
C) The temperature is normal after running
D) The blood pressure, after running, is normal for the patient

22. When marked fluid overload is accompanied by hemodialysis-dynamic instability, which of the following types of dialysate will best permit adequate ultrafiltration without hypotension?
A) High sodium dialysate
B) High potassium dialysate
C) High calcium dialysate
D) Acetate dialysate

23. Two and one half hours into the dialysis treatment, Mr. Smith experiences a drop in his blood pressure. The most common causes of hypotension during dialysis include all of the following EXCEPT:
A) Increased dietary intake of sodium
B) Taking antihypertensive medications before dialysis
C) Excess fluid removal
D) An unstable cardiovascular system

24. A low alarm in the pre-pump arterial pressure monitor may be the consequence of:
A) Line separation between patient and pump
B) Obstruction after the dialyzer
C) Decrease in blood pump speed
D) Hypotension or vasoconstriction

25. If a high venous pressure monitor alarm occurs during dialysis, you would first:
A) Check for an obstruction in the blood system, distal to the monitor site
B) Turn the blood pump off
C) Open alarm limits to reset
D) Clamp the venous needle

26. Fever and chills that develop midway through a patient's treatment are most likely caused by:
A) Pyrogenic contamination
B) Viral contamination
C) Infection
D) Incorrect dialysate temperature

27. The blood leak detector in most delivery systems:
A) Is a photoelectric cell placed in the effluent dialysis fluid circuit, which continuously measures the optical density of the fluid pumped through it
B) Is a photoelectric cell placed in the affluent dialysis fluid circuit, which continuously measures the optical density of the blood pumped through it
C) Is a photoelectric cell placed in the blood circuit, which continuously measures the optical density of the blood pumped through it
D) Is a cylinder placed in the dialysis in-flow line which continuously measures the density of the fluid pumped through it

28. The blood pump is set at 300 mL/min. The technician notices that the blood lines appear to be "jumping". The technician should:
A) Reposition the needle and adjust the blood flow downward because of excess negative pressure on the arterial portion of the circuit
B) Administer a bolus of saline because volume depletion has caused the arterial system to collapse
C) Turn off the pump and apply pressure because the needle has infiltrated
D) Administer normal saline because the patient is hypotensive

29. The major patient parameters in the urea kinetic model are:
A) Body urea volume (V)
B) Normalized protein catabolic rate (NPCR)
C) Residual renal urea clearance
D) All of the above

30. The treatment parameters in the urea kinetic model are:
A) Dialyzer urea clearance and treatment time
B) Ultrafiltration rate and dialyzer urea clearance
C) Treatment time and ultrafiltration rate
D) None of the above

31. You are ready to take Mr. K off dialysis. His pre-treatment temperature was 98.4° F and is now 101.4° F. Which physician's order will you anticipate?
A) Terminate dialysis stat
B) Draw blood for HBV, SGOT/SGPT
C) Administer an antipyretic and antibiotic
D) Draw blood for culture & sensitivity, CBC

32. Hemolysis during hemodialysis can occur as a result of dialyzing fluid that is heated to:
A) 43-44° C
B) 36-37° C
C) 39-40° C
D) 34-35° C

33. When a dialyzer membrane tears and allows blood to escape from the blood compartment into the dialyzing fluid compartment, it is termed:
A) A clotted dialyzer
B) A blood leak
C) Hemolysis
D) Blood line separation

34. What is the immediate treatment for the complication "hemolysis"?
A) Promptly stop the blood pump and clamp the blood lines
B) Promptly place the patient in Trendelenburg position and give oxygen via endotracheal tube
C) Promptly return hemolyzed blood and give a blood transfusion of packed cells
D) Promptly remove dialyzer and blood lines after taking the hematocrit

35. Acute hemolysis during hemodialysis is an extreme medical emergency. Clinical manifestations include:
A) Pain in vascular access, extremities, and/or chest, abdominal cramps, dyspnea, and arrhythmias
B) Headache, nausea, vomiting, cramping in extremities & chest pain
C) Hyperventilation, hypertension, disorientation & hallucinations
D) Weakness, nausea, vomiting, diarrhea & headache

36. What are the signs and symptoms of dialysis disequilibrium?
A) Headache, hypotension, confusion, and possible seizures
B) Headache, hypotension, muscle cramping and arrhythmias
C) Headache, hypertension, muscle cramping and nausea
D) Headache, hypertension, nausea and vomiting, and possible convulsions

37. Pyrogenic reactions are assumed to be the release of bacterial toxins across the dialyzer membrane to the patient. The source of the toxin may be from:
A) Improperly sterilized equipment or contaminated water supply for dialysate preparation
B) Improperly soaked dialyzers or contaminated water culture preparation
C) Improperly prepared dialysate or contaminated blood culture preparation
D) Improperly set temperature gauge or contaminated needles for dialysis initiation/preparation

38. An improperly functioning dialysate proportioning pump and conductivity monitor can result in the:
A) Delivery of a hypotonic dialysate resulting in acute hemolysis and water intoxication
B) Delivery of a hypertonic dialysate resulting in acute hemolysis and water intoxication
C) Delivery of an isotonic dialysate resulting in non-acute hemolysis and intoxication
D) Delivery of an isotonic dialysate resulting in semi-acute hemolysis and intoxication

39. After stopping an accidental air infusion in a patient undergoing hemodialysis, the practitioner's next course of action is to:
A) Turn the patient on his back and initiate cardiopulmonary resuscitation procedure
B) Position the patient on his left side with his head down and administer oxygen
C) Call for a bedside x-ray to determine the presence of air in the patient's heart
D) Position the patient on his right side with his head slightly elevated

40. Seizures during hemodialysis may be caused by all of the following <u>EXCEPT</u>:
A) Electrolyte imbalance
B) Hypotension
C) Sudden increase in dialyzer clearance
D) Dialysate composition errors

41. Hypotension during dialysis may be the result of:
A) Excessive or inaccurate volume depletion
B) Incorrect ultrafiltration rate
C) Antihypertensive medications
D) All of the above

42. The reason that potassium dialyzes out and red blood cells (RBC's) do not is because:
A) Potassium has a larger molecular weight
B) Potassium has a smaller molecular weight
C) RBC's are not water soluble
D) RBC's have a negative charge

43. The removal of urea from a patient in hemodialysis is primarily due to the existence of:
A) Osmotic pressure
B) Hydrostatic pressure
C) Electrical gradient
D) Concentration gradient

44. The purpose of using high dialysate flow rates during hemodialysis is to:
A) Maintain a high blood flow rate
B) Maintain a wide concentration gradient
C) Increase the dialysate bath temperature
D) Increase the pressure gradient

45. The following ion is usually present in the dialysis bath to correct the patient's metabolic acidosis?
A) Sodium
B) Chloride
C) Bicarbonate
D) Potassium

46. If you suspect an air embolism has occurred, what is (are) the appropriate actions to take after the blood pump is turned off and the blood lines are clamped?
I. Turn patient on his right side
II. Turn patient on his left side
III. Elevate head of bed rapidly
IV. Place patient in Trendelenburg position
A) I only
B) I and IV only
C) II and III only
D) II and IV only

47. Short frequent dialysis with slow urea removal when the BUN is very high is performed to prevent:
A) Rapid decreases in hematocrit
B) Dialysis disequilibrium syndrome
C) Cardiac arrhythmias
D) Excessive anticoagulation

48. A hypotonic dialysate bath will cause:
A) Hypernatremia
B) Hypotension
C) Vomiting
D) Hemolysis

49. Transmembrane pressure consists of which of the following pressure gradients on each side of the dialysis membrane?

I. Positive pressure on the blood side
II. Negative pressure on the blood side
III. Positive pressure on the dialysate side
IV. Negative pressure on the dialysate side

A) III and IV
B) I and IV
C) I, II and III
D) I, III and IV

50. The type of blood to dialysate flow used to produce an optimum gradient between blood and dialysate across the dialyzer membrane is:

A) Co-current flow
B) Cross-current flow
C) Counter-current flow
D) Parallel flow

51. Calculate the dialysate pressure that must be used to obtain a total TMP of 300mmHg pressure, using the following formula:

$$\frac{\textit{Pressure into dialyzer} + \textit{Pressure out of dialyzer}}{2} - \textit{Dialysate pressure}$$

Positive pressure into dialyzer = 100mmHg
Positive pressure out of dialyzer = 60mmHg

A) +140
B) -140
C) +220
D) -220

52. Factors that influence the effectiveness of solute removal during hemodialysis are:

A) Solute size, permeability of membrane, dialysate flow
B) Temperature of dialysate, blood flow rate
C) Concentration gradient, fluid removal, clotting
D) All of the above

53. Which of the following characteristics of a drug determine if it will be substantially removed by dialysis?

I. High molecular weight
II. Small volume of distribution
III. High degree of protein binding
IV. Water solubility

A) II and IV only
B) I, III and IV only
C) I and III only
D) All of the above

54. The correct formula to calculate mean transmembrane pressure is:
A) $[(P_{bi}+P_{bo})/2]-[(P_{di}+P_{do})/2]$
B) $[(P_{bi}-P_{bo})/2]-[(P_{di}-P_{do})/2]$
C) $(P_{bi}+P_{bo})-(P_{di}+P_{do})$
D) $(P_{bi}-P_{bo})-(P_{di}-P_{do})$

55. The dialysis machine assures the dialysate entering the dialyzer is safe by:
A) Regulating the temperature, conductivity, pH, and measuring pressure and flow
B) Alerting the user if something is wrong
C) Bypassing the dialyzer if dialysate is not safe
D) All of the above

56. The movement of water from an area of lower solute concentration to an area of higher solute concentration until equilibrium is reached is called:
A) Diffusion
B) Osmosis
C) Ultrafiltration
D) Dialysis

57. The potting compound, casing, fibers, and headers are part of a:
A) Dialysis machine
B) Water treatment system
C) Bicarbonate system
D) Dialyzer

58. The volume of plasma cleared of a given substance per unit of time is:
A) Clearance
B) Dialysis
C) Dialysance
D) Net flux

59. All of the following dialyzer membranes are considered to be biocompatible <u>EXCEPT</u>:
A) Polysulfone
B) Polyacrylonitrile
C) Polymethylmethacrylate
D) Cellulose acetate

60. Which of the following factors affect blood flow?

I. Needle size
II. Needle position
III. Stenosis
IV. Pump Speed

A) I and II
B) IV only
C) II, III, and IV
D) All of the above

61. Dialyzer compliance is:

A) The increase in dialyzer blood volume with decrease in TMP
B) The decrease in dialyzer blood volume with increase in TMP
C) The increase in dialyzer blood volume with increase in TMP
D) The decrease in dialyzer blood volume with decrease in TMP

62. The dialyzer with the least compliance is a:

A) Hollow fiber
B) Coil
C) Parallel plate
D) Cellulose acetate

63. A type of dialysate delivery system is a:

A) Batch system
B) Proportioning system
C) Regenerative system
D) All of the above

64. The following needs to be performed on a routine basis to ensure accurate blood pump occlusion:

A) Tightening of all nuts and bolts
B) Cleaning of pump header
C) Calibration
D) None of the above

65. If the dialysate fluid to be delivered to the dialyzer is < 34° C:

A) The patient's blood pressure may drop
B) The machine goes into bypass
C) The patient will complain of being cold
D) All of the above

66. If the dialysate temperature is > 41° C, the following will happen:

A) The machine goes into bypass
B) The venous line clamps
C) The blood pump stops
D) The blood pressure drops

67. Using sodium variation during hemodialysis may prevent this complication(s):
A) Hypotension
B) Cramping
C) Hypoalbuminemia
D) A & B

68. Glucose is used in dialysate during hemodialysis to:
A) Keep ions stable in solution
B) Prevent hypoglycemia in diabetic patients
C) Assist with regulation and removal of ions
D) All of the above

69. Long-term use of bio-incompatible membranes for hemodialysis is known to result in an increase in all EXCEPT:
A) Malignancy
B) Malnutrition
C) βeta 2 amyloid disease
D) Osteodystrophy

70. A possible factor(s) for obligatory weight loss when using non-volumetric machines is (are):
A) High venous pressure
B) High K_{UF}
C) Target weight loss is low
D) All of the above

71. When establishing a dry weight for the patient, the factor(s) that should be taken into consideration is (are):
A) Blood pressure
B) Patient well being
C) Evidence of dehydration or overload
D) All of the above

72. The process by which a large volume of fluid is removed at a rapid rate, with minimal solute removal is called:
A) Osmosis
B) Hemodialysis
C) Ultrafiltration
D) Isolated or pure ultrafiltration

73. The amount of solute leaving the blood entering the dialysate/unit of time is called:
A) Ultrafiltration
B) Dialysance
C) Net flux
D) Clearance

74. Factors that affect dialyzer clearance are:

I. Blood flow
II. Dialysate flow
III. Effective surface area
IV. Concentration gradient

A) I, III and IV only
B) I, II and III only
C) I and IV only
D) All of the above

75. The primary purpose of the proportioning pump in a dialysate delivery system is to:

A) Prepare the dialysate to the proper pH
B) Prepare the dialysate to the proper temperature
C) Prepare the dialysate to the proper water to concentrate ratio
D) Deliver the concentrate at the proper rate

76. When this complication occurs, the blood has a "cherry pop" appearance:

A) Hemolysis
B) Blood leak
C) Air embolism
D) None of the above

77. An early symptom of disequilibrium syndrome is:

A) Headache
B) Bradycardia
C) Hypotension
D) Euphoria

78. Prior to initiation of hemodialysis, the patient should be assessed for:

I. Fluid status
II. Blood pressure
III. Skin color, temperature, turgor & integrity
IV. Vascular access patency

A) I, II, III
B) I, II, IV
C) II, III, IV
D) I, II, III, IV

79. A high venous pressure alarm may be caused by which of the following?

    I. Needle placement
    II. Infiltration of the venous site
    III. Massive dialyzer blood leak
    IV. Clotting of the venous blood line

A) I and II only
B) I, II, and III only
C) I, II, and IV only
D) All of the above

80. Hemodialysis patients are discouraged from eating heavy meals before or during dialysis because it:

A) May cause hyperkalemia post dialysis
B) Can contribute to vomiting during dialysis
C) May contribute to hypotension
D) All of the above

81. A marker used to determine the middle molecule clearance of a dialyzer is:

A) Urea
B) Vitamin B_{12}
C) Nitrogen
D) Creatinine

82. Urea clearance is enhanced by:

A) High blood flow rate and a high dialysate flow rate
B) Co-current flow
C) A smaller dialyzer
D) Osmotic pressure gradient

83. The frequency of hospitalizations in ESRD patients has been contributed to:

A) Inadequacy of dialysis therapy
B) Low protein catabolic rates
C) Vascular access problems
D) All of the above

84. The delivery of the prescribed treatment plan can be compromised by:

    I. Wall clock syndrome
    II. Minimum UFR
    III. The technique used for obtaining blood specimens
    IV. Cutting treatment time

A) II and III only
B) I, III, and IV only
C) I, II, and III only
D) I and IV only

85. To minimize morbidity/mortality rates in the CKD Stage 5 patient population, the national standard for an adequate hemodialysis prescription (Kt/V) is:
A) ≥ 0.8
B) ≥ 0.4
C) ≥ 1.0
D) ≥ 1.2

86. Signs and symptoms of inadequate dialysis therapy may include:
I. Nausea and vomiting
II. Pericarditis
III. Increase in BUN and creatinine
IV. Hypotension
A) I and III only
B) I, II, and IV only
C) I, II, and III only
D) All of the above

87. The use of this should be avoided during the last hour of dialysis due to its effect on the movement of fluid from the interstitial spaces to the vascular compartment:
A) Epogen
B) Isolated ultrafiltration
C) Antibiotics
D) Hypertonic saline

88. Using a high sodium dialysate bath may predispose the patient to:
A) Fluid overload
B) Hypertension
C) Thirst
D) All of the above

89. The National Standard for the Urea Reduction Ratio (URR) is a ratio greater than:
A) 55%
B) 65%
C) 75%
D) 12%

90. The formula to calculate URR is:
A) $\frac{\text{PreBUN-PostBUN}}{\text{PreBUN}} \times 100 = \%\text{ URR}$
B) $\frac{\text{PreBUN+PostBUN}}{2} \times 100 = \%\text{ URR}$
C) $\frac{\text{PreBUN-PostBUN}}{\text{PostBUN}} \times 100 = \%\text{ URR}$
D) $\frac{\text{PreBUN+PostBUN}}{\text{PostBUN}} \times 100 = \%\text{ URR}$

91. Using the correct formula in question 90, calculate the URR for a patient with a PreBUN 88 and PostBUN 20.
A) 77.3%
B) 5.4%
C) 34%
D) 54%

92. In 1913, the artificial celluloid membrane and use of leech heads as the anticoagulant was developed by:
A) Willem Kolff
B) Scribner
C) Rowntree, Turner, and Abel
D) Teschan and Quinton

93. Willem Kolff developed the:
A) First disposable dialyzer
B) First plate dialyzer
C) Scribner shunt
D) Mahurker catheter

94. Hemolysis or inaccurate blood pump rates can result from:
A) Blood lines without an arterial drip chamber
B) Inaccurate calibration of the blood pump for lines used
C) Pediatric blood lines used for an adult patient
D) A poorly developed access

95. When one of the blood alarms is activated, the blood pump stops, there is an audio and visual alarm, and the venous line clamp closes. Alarms that would cause this to occur are:
A) Conductivity, pH, temperature, blood leak
B) Blood leak, venous pressure, temperature, conductivity
C) Arterial pressure, air detector, blood leak, venous pressure
D) Conductivity, arterial pressure, air detector, venous pressure

96. All of the following affect solute clearance <u>EXCEPT</u>:
A) Molecular weight of molecules
B) Blood flow rate
C) Dialyzer membrane
D) Type of access

97. The ability of a dialyzer to remove fluid is expressed as mL/hr/mmHg TMP. This is the dialyzer's:
A) Coefficient of ultrafiltration
B) Clearance
C) Surface area
D) Priming volume

98. Transmembrane pressure across the hemodialysis membrane consists of:

   I. Positive blood compartment pressure
   II. Negative dialysate compartment pressure
   III. Positive dialysate compartment pressure

A) I and II only
B) II and III only
C) I and III only
D) I, II and III

99. Pre-pump arterial pressure is:

A) The pressure required to pump the blood through the circuit
B) The resistance of the access to the blood flow out of the access
C) The pressure within the dialyzer
D) The pressure inside the access

100. The potential for hemolysis and vessel wall damage increases if the pre-pump arterial reading exceeds:

A) > +250mmHg
B) > -50mmHg
C) > -250mmHg
D) < +100mmHg

101. The characteristic of a dialyzer membrane which allows the blood volume to exceed the priming volume is called:

A) Surface area
B) Permeability
C) Compliance
D) Blood leak rate

102. A dialysate delivery system where concentrate and purified water are manually mixed in a reservoir and then delivered to the dialysis machines is called a:

A) Central delivery proportioning
B) Batch system
C) Regenerating system
D) Proportioning system

103. The delivery system where concentrate and water is mechanically mixed in a ratio, e.g 34:1, is a:

A) Proportioning System
B) Regenerative system
C) Batch system
D) Single pass system

104. Acetate baths are not commonly used today because they:
A) Harbor bacteria
B) Are very expensive
C) Can cause vasodilation
D) Can cause hypertension

105. Aluminum toxicity is a complication resulting from the use of :
A) Calcium acetate
B) Unpurified water
C) Amphogel
D) B and C

106. Signs and symptoms a patient may experience after an air embolism are:
A) Cyanosis, hypotension, SOB burning the access
B) Chest pain, dyspnea, coughing, cyanosis
C) Confusion, cherry pop blood, hypertension, double vision
D) Hypotension, double vision, SOB, hiccups

107. "Cherry pop" colored blood, hypotension, chest pain, and a drop in hemoglobin/hematocrit are signs of:
A) Residual chemical reaction
B) First use syndrome
C) Disequilibrium syndrome
D) Hemolysis

108. Bone disease is a long term complication in ESRD patients and is due to:
A) Metabolic acidosis
B) Poor compliance with PO_4^{---} binders
C) Ingestion of high phosphorus foods
D) All of the above

109. If hemolysis is suspected in a patient undergoing hemodialysis, it is important to do all <u>EXCEPT</u>:
A) Reinfuse blood, then clamp the bloodlines
B) Stop the blood pump and treat patient symptoms
C) Sample dialysate for pH & conductivity
D) Check patient's blood

110. The surface area of a hollow fiber dialyzer is determined by the:
A) Number of fibers
B) Inner diameter
C) Length
D) All of the above

111. According to the NKF-KDOQI guidelines, which statement is NOT true regarding the measuring and monitoring the delivered dose of hemodialysis?
A) Numerous studies have demonstrated a correlation between the delivered dose of hemodialysis and patient mortality and morbidity
B) Mortality among ESRD patients is lower when sufficient hemodialysis treatments are provided
C) Clinical signs and symptoms can be used along with laboratory data as indicators of hemodialysis adequacy
D) Morbidity and mortality among patients is higher when sufficient dialysis treaments are provided

112. The NKF-KDOQI guidelines recommend the delivered dose of hemodialysis using formal Urea Kinetic Modeling (UKM) employing the single-pool variable volume model because:
I. It provides the most accurate measurement of dialysis dose
II. It takes into account the contribution of ultrafiltration to the final delivered dose of dialysis
III. It is reproducible and quantitative
IV. It can be used to individualize a patient's dialysis prescription
A) I, II and III only
B) II, III and IV only
C) I, III and IV only
D) All of the above

113. A variety of factors may result in the actual delivered dose of hemodialysis falling below the prescribed dose, including:
I. Compromised urea clearance
II. Actual treatment time less than prescribed treatment time
III. Laboratory or blood sampling errors
IV. Residual renal function increase
A) I, II and IV only
B) I, II, and III only
C) II, III, and IV only
D) All of the above

114. NKF-KDOQI guidelines recommend the Slow Flow/Stop Pump technique to draw the post dialysis blood sample because:
I. It prevents sample dilution with recirculated blood
II. It minimizes the confounding effects of urea rebound
III. It is simple to implement in the clinical setting
IV. The Blood Reinfusion Sampling Technique has greater potential for increased dialysis prescription requirement
A) I, II, and III only
B) II, III, and IV only
C) I, II, and IV only
D) All of the above

115. Using the blood reinfusion sampling technique to obtain post-dialysis BUN samples may have Kt/V and URR values that are systematically lower than using the slow flow/stop pump technique.
A) True
B) False

116 The following action should be taken if the delivered Kt/V falls below1.2 or the URR declines to <65% on a single determination:
A) Investigate potential errors in the delivery of the prescribed hemodialysis
B) Empirically increase the prescribed dose of the hemodialysis
C) Suspend use of the reprocessed dialyzer
D) Dialyzer should be changed

117. A major barrier to providing an adequate hemodialysis prescription is:
A) Patient non-compliance
B) Intradialytic hypotension and/or cramps
C) Use of acetate dialysate
D) Use of reprocessed dialyzers

118. Reducing the dialysate temperature from 37° C to 34-35° C will result in:
I. Increased peripheral vasoconstriction
II. Increased cardiac output
III. Less hypotension
IV. Increased urea clearance
A) I, II, and III only
B) II, III, and IV only
C) III and IV only
D) All of the above

119. Treatment of hyperkalemia utilizing sodium bicarbonate and calcium gluconate may result in all of the following EXCEPT:
A) Temporarily antagonizes cardiac effects of hyperkalemia
B) May be indicated when acidosis is present
C) Temporarily translocates potassium into cells
D) The removal of potassium from the body in exchange for sodium

120. The long-term dialysis related complication of amyloidosis may manifest as all of the following EXCEPT:
A) Amyloid containing bone disease
B) Carpal tunnel syndrome
C) Amyloid arthropathy
D) Steal syndrome

121. Utilizing dialysate which is ultrapure may delay the onset of amyloidosis.
A) True
B) False

122. The purpose of rinsing a hemodialysis machine with an acid solution such as citric or vinegar is to:
   A) Disinfect the machine
   B) Prevent precipitate build up
   C) Rinse out chemicals
   D) All of the above

123. To prevent disequilibrium syndrome, it is important to rapidly lower the urea level by using a large dialyzer.
   A) True
   B) False

124. Techinque(s) used to treat or prevent muscle cramping during hemodialysis include all of the following <u>EXCEPT</u>:
   A) Sodium modeling
   B) A 50% dextrose solution
   C) Ultrafiltration modeling
   D) All of the above

125. Which of the following statements is <u>NOT</u> true regarding the conductivity of dialysate?
   A) Each ion type moves at a different speed in solution
   B) Different ion types have different charges
   C) Dextrose in the solution will increase the conductivity
   D) Some ingredients in dialysate do not ionize completely

126. The expected Kt/V of a patient whose treatment time is 4 hours, fluid volume is 40 L, and dialyzer clearance is 250 mL/min is:
   A) 1.05
   B) 1.25
   C) 1.50
   D) 1.55

127. The blood compartment pressure in a dialyzer is 250 mmHg and the TMP is 325mmHg. The dialysate pressure is:
   A) 575 mmHg
   B) 75 mmHg
   C) –75 mmHg
   D) –575mmHg

128. Characteristics of a solute which promote its removal by dialysis are:

   I. Small molecular size
   II. Small volume of distribution
   III. Water solubility
   IV. Low protein binding

A) I, II, and III only
B) I and II only
C) II, III, and IV only
D) All of the above

129. All of the following statements concerning "first use syndrome" are true EXCEPT:

A) This is an allergic reaction to new dialyzers
B) Back pain, chest pain, and difficulty breathing may be manifested
C) Symptoms are usually manifested within 15 minutes of contact
D) Synthetic membranes are more commonly associated with this syndrome

130. Which of the following statements is/are true regarding disequilibrium syndrome?

   I. Most common in severely catabolic states
   II. May be manifested by headaches, confusion, and seizures
   III. An occurrence related to cerebral edema
   IV. Can only be seen in a patient who has never had dialysis before

A) I only
B) I and IV only
C) I, II, and III only
D) All of the above

131. According to the National Kidney Foundation's Kidney Disease Outcome Quality Initiative (KDOQI) 2006 guidelines, patients with thrice-weekly dialysis may experience an increase in mortality when their:

A) Spkt/V increases to 2.4
B) Spkt/V falls below 1.2
C) URR falls to 40%
D) Spkt/V has a value of 1.25

132. When attempting to minimize blood loss during hemodialysis therapy, the following should be done:

   I. Pretesting dialyzers
   II. Monitoring heparinization
   III. Assuring complete rinse back of blood and minimizing blood sampling
   IV. Regular monitoring and maintenance of dialysis equipment

A) I and II only
B) II and III only
C) II, III and IV only
D) All of the above

133. During hemodialysis, replenishing the dialysate compartment with fresh dialysate and the blood compartment with undialyzed blood, the concentration gradient is being widened and:
A) Preventing equilibrium
B) Maximizing clearance
C) Facilitating diffusion
D) All of the above

134. The removal of urea during dialysis is not highly influenced by the erythrocyte count because it diffuses easily from the red blood cells to the plasma.
A) True
B) False

135. Creatinine and phosphorus are found in red blood cells and plasma at the same levels and therefore diffuse easily during dialysis.
A) True
B) False

136. A dialyzer with a higher KoA is needed when the blood flow rate increases and the level of urea is measured at the dialyzer outlet:
A) Increases
B) Decreases
C) Normalizes
D) The level of urea is unchanged

137. During hemodialysis, creatinine clearance is approximately 80% of the level of urea clearance.
A) True
B) False

138. For most patients, urea rebound is nearly complete 15 minutes after dialysis.
A) True
B) False

139. The acid storage tanks for concentrate should be:
A) Clearly marked
B) Tested regularly for proper content
C) Drained and discarded periodically
D) A and B

140. Dialysis machine drains require:
A) Direct connection to drain waste vent (DWV)
B) A torturous path connection
C) A 1" air gap
D) A metallic basin

141. Kinetic modeling is used for all of the following <u>EXCEPT</u>:
A) Monitoring patient care
B) Measuring efficiency of dialysis
C) Determining dialysate composition
D) Individualizing dialysis prescriptions

142. The most effective way to increase ultrafiltration during hemodialysis is by increasing the:
A) Dialysate flow rate
B) Membrane surface area
C) TMP
D) Dialysate dextrose in excess of 200mg %

143. The transmembrane pressure has been gradually increasing over the last hour of treatment. The practitioner suspects a clotted dialyzer. The first action the practitioner should take is:
A) Check for a decreased arterial flow
B) Assess for a clotted venous needle
C) Change the clotted dialyzer
D) Flush the extracorporeal circuit with normal saline

144. During dialysis the air detector alarms, and foam is visible in the venous line. The immediate action of the practitioner should be to:
A) Mute the alarm
B) Hit the venous chamber
C) Clamp the venous line
D) Continue the treatment

145. A patient is taking digoxin 0.125mg every day. An order is given to place the patient on 3.0 mEq/L potassium bath. The rationale for this dialysate prescription is:
A) To prevent digoxin toxicity, the rate of potassium removal is more important than the amount
B) To prevent digoxin toxicity, the amount of potassium removed is more important than the rate
C) The clearance rate of digoxin potentiates the action of potassium
D) The clearance rate of digoxin exceeds that of potassium

146. A patient is gaining 3 kgs and more between treatments. To help control the patient's weight gains, the best suggestion by the practitioner would be to:
A) Eat more fruits and vegetables and less meat and carbohydrates
B) Eat ice cream instead of drinking a soda between meals
C) Suck on hard candies or ice, or chew gum to relieve a dry mouth
D) Take no liquids after 6pm on the day before dialysis

147. If air is detected in the venous line, the air detector should simultaneously perform which of the following functions?
A) Clamp the venous line, auditory and visual alarms and stop the blood pump
B) Auditory alarm, reduces the venous pressure and stops the blood pump
C) Dialysate goes into bypass and the machine stops
D) Auditory and visual alarm and blood pump stops

148. As you are returning blood to the patient at a slow rate at the end of dialysis, you notice that the venous chamber has emptied and that the venous line has filled with air. Your FIRST response should be to:
A) Clamp the venous line and turn off the blood pump
B) Clamp the venous line and leave on the blood pump to fill the venous drip chamber
C) Disconnect the patient from the venous line, remove air, prime the venous line and continue with the procedure
D) Lay the patient flat on the left side

149. If the patient's dry weight is set too high, the patient will:
A) Become dehydrated and require additional saline infusion
B) Remain in a fluid-overloaded state at the end of the dialysis session
C) Experience frequent hypotensive episodes during the latter part of the dialysis session
D) Lose weight and experience a rapid rebound in the blood urea nitrogen level

150. The MOST COMMON cause(s) of death among hemodialysis patients is (are):
A) Gastrointestinal bleeding
B) Sustained elevation of BUN and creatinine
C) Cardiac complications
D) Dialysis dementia

151. In selecting a dialyzer for a hemodialysis patient, which of the following is/are LEAST important?
A) Patient's age
B) Ultrafiltration coefficient
C) BUN and creatinine clearances
D) Priming volume

152. All of the following are potential problems associated with the use of bicarbonate dialysate EXCEPT:
A) Precipitation of calcium and magnesium in the delivery system
B) Carbon dioxide levels in the bicarbonate dialysate becoming dangerously high
C) Increased risk of bacterial growth
D) Dangerous electrolyte concentrations due to concentrate mixups

153. The total pressure in the blood circuit is a result of all of the following EXCEPT the:
A) Dialysate flow rate
B) Blood flow rate controlled by the blood pump
C) Resistance to flow in the blood circuit
D) Internal diameter of the fistula needles

154. Hyperkalemia is a severe complication for patients on hemodialysis because it can produce:
A) Intracellular fluid loss
B) Lethal cardiac arrhythmias
C) Extracellular fluid loss
D) Anorexia

155. The dextrose concentration of dialysate does not add to the total ionic content because it:
A) Has catatonic charges that cannot be measured
B) Only ionizes when concentrated
C) Is not ionized
D) Has a positive electric charge

156. The primary function of a proportioning pump in a dialysis delivery system is to:
A) Reduce the pressure of the water inflow from the tap connection
B) Continuously measure and mix the water and concentrate to form dialysate
C) De-ionize the concentrate and the dialysate
D) Separate the air from the dialysate before it reaches the bubble trap

157. Which of the following descriptions best defines diffusion?
A) Water movement across a membrane in response to a hydrostatic pressure differential
B) Water molecules moving from an area of high concentration to an area of low concentration until equilibrium is reached
C) Water molecules moving from an area of low concentration to an area of high concentration until equilibrium is reached
D) Particles, in solution, moving from an area of high concentration to an area of low concentration until equilibrium is reached

158. To prevent hemolysis, the thermal sensor and regulator of the dialysate delivery system should be maintained to keep the dialysate within what temperature range?
A) 35° C +/- 10° C
B) 37° C +/- 5° C
C) 42° C +/- 2° C
D) 37° C +/- 1° C

159. A patient complains of excruciating pain at the site of the venous needle when dialysis is begun with a re-used dialyzer. Your FIRST response should be to:
A) Insert another venous catheter
B) Turn down the blood pump
C) Clamp the venous line and stop the blood pump
D) Put the patient in trendelenburg position on the left side

160. During dialysis, an increase in the blood circuit pressure can be caused by all of the following EXCEPT:
A) A clotted dialyzer
B) Increased arterial flow
C) Venous needle infiltration
D) Excessive ultrafiltration

161. A practitioner notices that a patient's blood has turned very dark. He/she draws samples from the arterial and venous lines and a peripheral vein in order to:
A) Determine the KOA of the dialyzer
B) Calculate recirculation in the access
C) Measure the treatment Kt/V value
D) Assess the potassium level as an indication of hemodialysis

162. All conductivity meters function based on their ability to:
A) Pass an electrical current through an electrolyte solution
B) Measure the viscosity of a fluid
C) Compare a solution's density to its temperature
D) Create gas bubbles at its electrodes

163. A hemodialysis patient who has volume-dependent hypertension will generally be able to control his blood pressure by:
A) Observing a moderate sodium restriction in his diet, getting sufficient rest and decreasing worry
B) Taking a diuretic, which will deplete his body's water volume and reduce pressure in the blood vessels
C) Drinking whenever he is thirsty, which is the most reliable means of assuring that his body has the proper amount of fluids
D) Maintaining an adequate fluid balance and following a proper sodium restriction in his diet

164. In response to a high temperature alarm, your first action should be to:
A) Replace the machine
B) Notify the machine technologist
C) Verify that dialysate flow is stopped
D) Take the patient's temperature

165. Calculate the TMP for a 3 kg weight loss on a hollow fiber dialyzer with a KUF of 4.0. Assume a four hour treatment with 1000cc replacement in prime, rinseback, and fluid intake. The venous resistance at QB=300 is 180 mmHg:
A) Approximately 250, 70 mmHg negative pressure
B) Approximately 1000, no negative pressure
C) Approximately 100, 150 mmHg negative pressure
D) Approximately 200, 20 mmHg negative pressure

166. A high venous pressure alarm may occur in all of the following circumstances EXCEPT:
A) Infiltrated venous return
B) Dislodged arterial needle
C) Needle resting on the side of the vein
D) Kinked venous line

167. In hemodialysis, 0.9%NaCl is used. If this solution has the same effective osmolality as body fluids, it is called:
A) Hypertonic
B) Isotonic
C) Hyptonic
D) Monotonic

168. Hemodialysis is accomplished by which combination of processes?
A) Ultrafiltration, filtration, osmosis
B) Osmosis, filtration, diffusion,
C) Filtration, osmosis, dilution
D) Osmosis, diffusion, ultrafiltration

169. When assessing the adequacy of dialysis, which of the following factors need to be considered?
I. Metabolic state
II. Protein intake
III. Hydration
IV. Height and weight
A) I and II only
B) II only
C) I, II, and III only
D) All of the above

170. In the expression Kt/V, the "K" stands for:
A) Serum potassium
B) Urea clearance of the dialyzer
C) Kidney
D) Kinetic modeling

171. All of the following are considered contributing factors to the occurrence of muscle cramps during hemodialysis treatment EXCEPT:
A) Hypoglycemia
B) Hypovolemia
C) Hyperosmolality
D) Hyponatremia

172. The Association for the Advancement of Medical Instrumentation (AAMI) Standards recommend that dialysate pH be kept above 6.0 because:
A) Acetate precipitates in the presence of the calcium below this value
B) Potassium concentration may elevate when the pH is below this value
C) Heparin inactivation has been observed below this value
D) Bicarbonate causes breakdown of polyvinyl chloride releasing toxic substances into the bloodstream

173. Some of the safety parameters for the dialysate on the dialysis machine are:
A) Regulating the temperature, conductivity, pH
B) Measuring the pressure and flow rate
C) By passing the dialyzer, if the dialysate is not safe
D) All of the above

174. Pyrogenic reactions are assumed to be the release of bacterial toxins across the dialyzer membrane to the patient. The source of the toxins may be from:
A) Improperly sterilized equipment or contaminated water supply for dialysate preparation
B) Improperly soaked dialyzers or contaminated water culture preparation
C) Improperly prepared dialysate or contaminated blood culture preparation
D) Improperly set temperature gauge or contaminated needles for dialysis

175. A hemodialysis machine indicates a high venous alarm and the blood pump has stopped. The practitioner should do which of the following first?
A) Insert a new venous needle
B) Check patient's blood pressure
C) Reset the alarm and increase blood flow
D) Check for an obstruction in the venous line

176. The storage time of aqueous bicarbonate concentrate is limited because:
A) Bacterial growth may occur
B) Of degassing of CO_2 solution
C) Old bicarbonate will affect ultrafiltration
D) Bicarbonate growth causes leaching in the container

177. The patient's blood pressure may begin to fall if you are pulling too much water too quickly from the:
A) Interstitial space
B) Intracellular space
C) Vascular compartment
D) Tissues

178. Fever during dialysis may be caused by:
A) Hyponatremia
B) Infection
C) Introduction of pyrogen or endotoxin during dialysis
D) B and C

179. Hemolysis during dialysis is related to exposure of the patient's blood to:
A) Hypotonic solutions
B) Choloramines and nitrosamines
C) Formaldehyde
D) All of the above

180. Dialysis encephalopathy is thought to be related to:
A) Ingestion of aluminum in phosphate binding medications
B) Absorption of aluminum from water used to prepare dialysate
C) Intake of aluminum from blood transfusion
D) A and B

181. Dialysis disequilibrium syndrome may be prevented by all of the following EXCEPT:
A) Slow blood flow rate
B) Short, frequent dialysis
C) High ultrafiltration rate
D) Administration of an osmotic agent

182. What symptoms might be manifested in the patient experiencing an air embolism?
A) Chest pain, shortness of breath, cough
B) Chest pain, shortness of breath, burning in the access
C) Confusion, cherry pop colored blood
D) Hypotension, double vision, access pain

183. The most common cause of hypotension during dialysis is:
A) The removal of sodium and water from the intravascular space by ultrafiltration
B) The removal of sodium and water from the intravascular space by osmosis
C) The removal of sodium and water from the intravascular space by diffusion
D) The removal of potassium and water from the intravascular space by osmosis

184. Treatment for hypotension includes:
A) Increasing the UFR and administering fluid
B) Administering IV fluids and placing the patient in the Trendelenburg position
C) Placing the patient in Trendelenburg position and increasing the UFR
D) Placing the patient in upright position and decreasing the TMP

185. To assess Mr. Burn's fluid volume status, the clinician will look at the following parameters:
A) Presence of peripheral edema, rapid pulse and uremic odor
B) Presence of edema, neck vein distention and abnormal breath sounds
C) Presence of azotemia, ascites and irregular heartbeat
D) Presence of ascites, pale color and cyanotic nail beds

186. In response to a patient's concern about a 2kg weight gain between dialysis treatments, the practitioner should make which of the following interpretations?
A) Even a moderate weight gain can mean salt overload
B) A moderate weight gain will stimulate the kidneys to resume urine production
C) A 1-2kg gain in weight is undesirable and could result in hypertension
D) A gain of 1-2kg in adults is safe and will help maintain a normal blood pressure

187. If the venous needle is inadvertently pulled out:
A) There will be a progressive rise in pressure at the post pump monitor
B) The line will automatically become kinked, thereby turning the pump off
C) A catastropic hemorrhage may result, since the outflow pressure will not increase and the pump will continue to operate
D) A message will be sent to the air detector that will trigger an alarm and then shut off the power supply

188. Potential causes of a hypotensive episode during hemodialysis include:
A) Hypoxemia
B) Medications
C) Eating
D) All of the above

189. All of the following situations would cause a decrease in venous pressure, <u>EXCEPT</u>:
A) Obstruction between the arterial drip chamber and the dialyzer
B) Obstruction between the patient's access and the arterial drip chamber
C) Obstruction between the venous drip chamber and the patient's access
D) An obstruction between the dialyzer and the venous chamber

190. Mrs. M., who is a 72-year old widow, asks to have her treatment shortened by 20 minutes today so she can get home and watch her favorite soap opera. You explain that:
A) This will be okay because a few minutes here and there doesn't make any difference
B) Unit policy does not permit patients to discontinue treatment early
C) It is very important to get the prescribed time every dialysis treatment to avoid medical complications
D) She can leave, but she must sign a release

191. Bob S. is a retired postal worker, who is on dialysis because of hypertension. Bob has been to a party and comes to dialysis with 6kg of fluid to be removed. You plan his treatment with the nurse and also:
A) Deliver a stern lecture to Bob about his overindulgence
B) Remind Bob that beer is contraindicated because of the blood pressure medications he is on, and threaten to notify the physician
C) Review the foods and fluids Bob has consumed and help him to see where he could have made better choices
D) Avoid mentioning Bob's excess weight gain. You don't want to upset him.

192. In planning for Bob's care, you will be alert for symptoms that the treatment is removing too much fluid too quickly. These symptoms include:
A) Low blood pressure, cramps, dizziness
B) Shortness of breath, fatigue, restlessness
C) Fever, headache, tremors
D) High blood pressure, swelling, cough

193. Mrs. K. is new to dialysis. She is very unsure about what is going on about her and she asks you some questions about her dialysis. She doesn't understand why her blood pressure is checked during dialysis. You tell her that:
A) You are monitoring her blood pressure because that is one way to make sure you are not removing too much fluid from her during dialysis
B) Dialysis is a pretty dangerous procedure to do on anyone and it is important that you record her blood pressure
C) Dialysis is a procedure that removes waste products during dialysis and blood pressure is one way to measure how much waste is being removed
D) Dialysis works primarily by applying diffusion and osmosis; blood pressure measures those principles

194. Each of the following is a sign of fluid overload <u>EXCEPT</u>:
A) Puffy eyelids
B) Numbness and tingling in the hands and feet
C) Distended neck veins
D) Shortness of breath

195. Which of the following substances is found in the dialysate?
A) Magnesium
B) Urea
C) Urate
D) Phosphate

196. Ultrafiltration is:
A) The measure of solute movement during dialysis
B) The process used to remove excess fluid during dialysis
C) Dependent on the chemical composition of the dialysis fluid
D) Dependent on the dialysate and blood component pressures in the dialyzer

197. To assure accurate blood flow, the following must be performed periodically:
A) Tightening of all nuts and bolts
B) Cleaning of the pump header
C) Calibration of the pump
D) Replacement of the pump

198. Factors affecting the rate of diffusion in hemodialysis include:
A) Molecular size and membrane pore size distribution
B) Surface area of the dialyzer
C) Concentration gradient between blood and dialysate
D) All of the above

199. Resistance to diffusion includes all of the following components <u>EXCEPT</u>:
A) Blood film layer or thickness
B) Membrane
C) Dialysate film/thickness
D) Solute drag

200. Factors affecting net flux include:
A) Membrane surface area
B) Membrane permeability
C) Blood-dialysate flow configuration
D) All of the above

201. Clearance is an expression of:
A) The performance of the dialyzing process
B) The volume of blood totally cleared of a given solute per unit of time
C) The blood flux rate per minute
D) Both A & B

202. Ultrafiltration of fluid from blood to dialysate occurs because of:
A) Dialysate flow rate
B) Hydrostatic pressure gradient between blood and dialysate
C) Blood solute concentration
D) None of the above

203. The roller-type blood pump, the two important parameters which are most important for blood flow rates are:
A) Pump speed and resistance
B) Pump speed and diameter
C) Correct occlusion and routine calibration
D) Vascular resistance and routine calibration

204. Signs and symptoms of hemolysis include all of the following <u>EXCEPT</u>:
A) Clear, cherry red blood in the venous line
B) Hypotension
C) Diarrhea and vomiting
D) Chest pain and dyspnea

Section VI
VASCULAR ACCESS

1. The first permanent type access, the shunt, was developed by:
   A) Scribner and Quinton
   B) Kolff
   C) Brescia and Cimino
   D) Turner

2. An access in which there is a synthetic type material placed subcutaneously to join an artery and a vein is called a:
   A) Shunt
   B) Fistula
   C) Graft
   D) None of the above

3. The fistula was developed by:
   A) Turner
   B) Sribner
   C) Quinton
   D) Brescia and Cimino

4. The correct angle in which to cannulate a fistula is:
   A) 10-15 degree angle
   B) 25-35 degree angle
   C) 35-45 degree angle
   D) No specific angle

5. Factor(s) that affect access recirculation is (are):
   A) Distance between arterial to venous needle tips
   B) Venous pressure
   C) Direction of the needle tips
   D) All of the above

6. The formula for calculating access recirculation is:
   A) (Arterial – Venous) / (Peripheral - Venous) x 100
   B) (Peripheral – Venous) / (Peripheral – Arterial) x 100
   C) (Arterial – Venous) / 2 x 100
   D) (Peripheral – Arterial) / (Peripheral – Venous) x 100

7. An acceptable percentage of access recirculation in a fistula is:
   A) < 10%
   B) < 18%
   C) < 25%
   D) < 50%

8. In a mature AV fistula or graft, what size of needle is needed to support the required blood flow rate for high-efficiency dialysis?
   A) 21 gauge
   B) 19 gauge
   C) 18 gauge
   D) 15 gauge

9. AV fistulas should be mature by:
   A) 2 weeks
   B) 12 weeks
   C) 6 weeks
   A) None of the above

10. An internal access is clotted if there is:
    A) The presence of thrill
    B) An absence of a thrill
    C) The presence of bruit
    D) Poor blood flow

11. The purpose of a chest x-ray after the insertion of a permanent catheter is to:
    A) Confirm proper placement of the catheter
    B) Confirm patency of the catheter
    C) Confirm patency and position
    D) Confirm position and absence of pneumothorax

12. Shunts are no longer used as a permanent access for hemodialysis due to the:
    A) High incidence of infection
    B) Potential for hemorrhage
    C) Use of fistulas and grafts
    D) All of the above

13. A disadvantage of an AV fistula over an AV graft is:
    A) The required length of time for maturity to take place
    B) A high incidence of infection
    C) A high incidence of thrombosis
    D) The decreased life span of the access

14. Exercise for the development of an AV fistula should:
    A) Begin 10-14 days post operative
    B) Take place after swelling of the anastomosis has diminished
    C) Be performed 3-4 times a day, for 5-10 minutes
    D) All of the above

15. Proper maturation of an AV fistula:
    A) Requires a minimum of two weeks
    B) Is necessary to avoid infiltration on cannulation
    C) Will enhance the life of the access
    D) B and C

16. Thrombosis and subsequent loss of an AV access is attributed to:
    A) Frequent hypotensive episodes
    B) Infections
    C) Stenosis
    D) All of the above

17. Signs and symptoms of venous stenosis include:
    A) Recirculation >10%
    B) High intradialytic venous pressure
    C) Increase in midweek BUN
    D) All of the above

18. A major cause of aneurysms and pseudoaneurysms in an AV access is:
    A) Infection
    B) Inappropriate use of tourniquets
    C) Not rotating sites
    D) Recirculation

19. All of the following are true when cannulating an AV fistula <u>EXCEPT</u>:
    A) Needles should be placed at least 1-2 inches apart
    B) Using a tourniquet
    C) Using a 45-90 degree angle
    D) Practicing good aseptic technique

20. Your patient complains of mild to moderate pain in the access limb. The limb is noted to be cooler distally and paler than the other limb. The patient states that the pain increases in intensity while he is on dialysis. The patient could be experiencing:
    A) Recirculation
    B) Infection
    C) Steal syndrome
    D) Black blood syndrome

21. Intervention for the above complaint in Question #20 would include:
    A) Placing the affected extremity in a dependent position and keeping it warm
    B) Elevating the limb on a pillow and keeping it warm
    C) Applying cold compresses to reduce swelling
    D) Drawing recirculation blood

22. Causes of an infection in an AV graft include:
A) Break in aseptic technique during cannulation
B) Poor hygiene and care of the access arm
C) Bacterial seeding from another infected site in the body
D) All of the above

23. While assessing an access before cannulation you note an aneurysm. You should:
I. Alert the physician
II. Avoid cannulation of area
III. Not cannulate the access
IV. Note the size of the aneurysm
A) I and II only
B) I, II, and IV only
C) I and III only
D) All of the above

24. Patient teaching concerning an aneurysm includes:
I. Avoiding trauma to the area
II. Wrapping the access at all times when not on dialysis
III. Reporting any signs of infection
IV. Encouraging rotation of sites
A) I, III, and IV only
B) I and II only
C) III and IV only
D) All of the above

25. Cannulation of an AV graft is best accomplished by using a 45 degree angle of the needle because it:
A) Avoids destruction along the surface of the AV graft
B) Steeper angles don't allow sufficient blood flow
C) Coring will occur at angles <45 degrees
D) Prevents the formation of an aneurysm

26. According to NKF DOQI guidelines, what is the correct order for the first and subsequent placement of vascular access?
I. A wrist (radial-cephalic) primary AV fistula
II. An elbow (brachial-cephalic) primary AV fistula
III. An arteriovenous graft of synthetic material
IV. A transposed brachial-basilic vein fistula
A) I, II, III, IV
B) II, IV, III, I
C) IV, III, II, I
D) III, II, I, IV

27. Subclavian vein catheterization should be avoided for temporary access due to the increased risk of:
A) Central vein stenosis
B) Pneumothorax
C) Infection
D) Difficult insertion

28. Referral for placement of permanent vascular access should be initiated when:
A) The creatinine clearance is < 25 mL/min
B) Serum creatinine level is > 4 mg/dL
C) Within 1 year of the anticipated need for dialysis
D) When the patient begins to show symptomatology

29. A new primary arteriovenous fistula should be allowed to mature before cannulation for at least:
A) 12 weeks
B) 6 weeks
C) 2 weeks
D) 10 weeks

30. The risk of access thrombosis increases when:
A) The blood pump speed rate is lower than 600 mL/min
B) The graft blood flow is < 600 mL/min
C) An access requires more than 5 minutes of compression to establish hemostasis
D) If the patient is on coumadin

31. Access recirculation studies are a relatively late predictor of venous stenosis.
A) True
B) False

32. Measurement of recirculation is more useful in patients with AV fistula than those with grafts.
A) True
B) False

33. The three needle peripheral vein method of measuring recirculation should <u>NOT</u> be used because it:
I. Overestimates access recirculation
II. Requires unnecessary venipuncture
III. Underestimates the peripheral urea value
IV. Gives a falsely elevated venous urea value
A) I, II, and III only
B) I and II only
C) I and III only
D) All of the above

34. Venous stenosis contributes greatly to morbidity rates in hemodialysis patients because it:

    I. Increases the risk of thrombosis
    II. Increases the risk of infection
    III. Decreases the need for heparin
    IV. Reduces the efficiency of the hemodialysis treatment

    A) I and II only
    B) I and IV only
    C) I, III, and IV only
    D) All of the above

35. Degenerative changes within a graft or the overlying skin may lead to:
    A) Compromised circulation
    B) Incomplete hemostasis on needle withdrawal
    C) Graft rupture
    D) All of the above

36. A primary AV fistula using the cephalic vein affords the best permanent access with the fewest complications.
    A) True
    B) False

37. The use of cuffed catheters for chronic hemodialysis is associated with:
    A) Lower blood flow rates
    B) Increased risk of infections
    C) Patient reluctance to consider more permanent access options
    D) All of the above

38. Which fibrinolytic agent is the only currently available FDA approved drug for clearing thrombolytic occlusions in central venous catheters?
    A) Urokinase
    B) Streptokinase
    C) Alteplase
    D) Reteplase

39. Arterial stenosis in an access is suspected if there is:
    A) A less negative arterial reading
    B) A more negative arterial reading
    C) A decrease in circulation
    D) An increase in venous pressure

40. Complications associated with arteriovenous fistula for hemodialysis access include which of the following:
A) Thrombosis, infection, stenosis, arterial steal syndrome
B) Infection, aneurysm, hypovolemia, epistaxis
C) Stenosis, thrombosis, hypovolemia, hypoxia
D) Arterial steal syndrome, infection, epistaxis, hyperthalemia

41. Infection is a serious complication in a graft because:
A) There is a danger of graft disintegration and subsequent hemorrhage
B) The graft will thrombose
C) The graft will form a stenosis at the anastomosis
D) An aneurysm will probably form at puncture sites

42. E.T. is a 62-year-old female on hemodialysis. She has a left upper upper arm A-V graft that has functioned well for two years. Recently, there has been suspicion that the percent recirculation in the graft may be rising. The suspicion is based on the observation of:
A) Color of the distal extremity
B) Increased heparin requirements
C) Unexplained increases in BUN and creatinine
D) Frequent clotting in the dialyzer

43. In order to determine which end of a graft is anastomosed to an artery in a patient, the practitioner should:
A) Aspirate blood and observe its color
B) Apply a tourniquet above the graft and observe for arterial pulsation
C) Insert a fistula needle pointing toward the heart and observe the blood flow after opening the clamp
D) Press firmly with the two middle fingers at the midpoint of the graft and palpate for the thrill

44. The practitioner should use which of the following actions to prevent venous aneurysms and thrombosis in a fistula?
A) Pre-dialysis scrub of the fistula
B) Elevation of extremity post-dialysis
C) Administration of systemic anticoagulants
D) Rotation of puncture sites

45. Correct treatment for a hematoma includes:
A) Warm packs during dialysis
B) Ice packs during dialysis
C) Heavy, direct pressure over the affected area
D) Immediate surgical intervention

46. One complication associated with a graft access device is:
A) Too rapid venous flow
B) Bleeding at puncture sites on non-dialysis days
C) Aneurysm formation mid-graft
D) Stenosis at the venous anastomosis

47. Aneurysm formation:
A) Generally poses no problems unless it becomes thin walled
B) Needs to be repaired immediately
C) Can occur at any time and become life threatening
D) Tends to clot easily

48. When a needle is being removed from an arteriovenous fistula hemostasis is BEST achieved by:
A) Moderate proximal pressure on the puncture sites
B) Heavy direct pressure over the puncture sites
C) Applying proximal pressure with a finger between the venous puncture site and the venous anastomosis
D) Applying a tourniquet for 10 minutes directly over the puncture site.

49. Mr. Brown, a 58-year-old male, was started on hemodialysis emergently with a cuffed venous catheter. A chest x-ray was ordered post catheter insertion. An immediate complication, post insertion of a catheter that can be detected by x-ray is:
A) Infection
B) Pneumothorax
C) Thrombus
D) Dysrhythmia

50. When assessing a tunneled catheter, you should inspect and palpate for signs of infection at the:
A) Synthetic cuff
B) Arterial lumen
C) Venous lumen
D) Exit site

51. Signs and symptoms of catheter malposition in the jugular vein include:
A) Swollen arm on the side of the catheter
B) Spike in temperature
C) Sudden ear ache on the side of the catheter
D) Sudden increase in heart rate

52. Which intervention may correct catheter malfunction related to occlusion by a fibrin sheath?
    A) Thrombolytic therapy with TPA
    B) Infusing an antibiotic
    C) Pulling the catheter back slightly
    D) Changing the dressing and catheter cap

53. The patency of a catheter is usually maintained after dialysis by:
    A) Irrigating with normal saline
    B) Filling the catheter with heparin
    C) Infusing the prescribed amount of TPA
    D) Avoiding catheter use until the next dialysis

54. According to the *NKF/DOQI Clinical Practice Guidelines on Vascular Access*, quality of life and outcomes in vascular access could be improved with:
    A) Increased use of cuffed tunneled catheters and decreased use of AV fistulae
    B) Increased use of newer catheters that have flow rates comparable to fistulae
    C) Increased use of AV fistulae and detecting dysfunction prior to thrombosis
    D) Increased use of radiographic imaging to confirm thrombotic occlusion

55. According to K/DOQI Vascular Access guidelines, the preferred site of insertion for a central venous catheter is:
    A) Superior vena cava
    B) Right internal jugular vein
    C) Left external jugular vein
    D) Inferior vena cava

56. Which type of hemodialysis vascular access comes closest to fulfilling the criteria of an "ideal" vascular access?
    A) AV fistula
    B) Cuffed tunneled central venous catheter
    C) Prosthetic graft
    D) AV graft

57. Which of the following statements about maturation of an AV fistula is <u>NOT</u> true?
    A) Maturation describes the process whereby an artery increases in size and wall thickness over time
    B) Maturation occurs as a result of increased pressure flow and pressure within the vein
    C) An AV fistula can take from 4 weeks to several months to fully mature
    D) If the AV fistula does not appear to be maturing after 4-6 weeks, an interventional radiologist should be consulted

58. The two most common complications associated with central venous access devices are:
A) Catheter related occlusion
B) Catheter related infection
C) Inadvertent removal of the catheter by the patient
D) A and B

59. Which one of the following observations would indicate a clotted fistula or graft?
A) Absence of a bruit/thrill
B) Numbness of the associated limb
C) Swelling of surrounding area
D) Unusual redness

60. With central venous catheters, poor blood flow can be caused by all of the following EXCEPT:
A) Blood pumps set higher than 200 ml/min
B) Incorrect placement of the cannula within the host vessel
C) Vascular spasms
D) An incorrect cannula lumen size

61. Which one of the following is among the complications associated with an infected access?
A) Bacterial endocarditis
B) Loss of the limb
C) Reduced vascular flow
D) Venous stenosis

62. Recirculation is the mixing of outflow blood with inflow blood, leading to inadequate dialysis. All of the following conditions will contribute to recirculation during dialysis EXCEPT:
A) An access having a low blood flow rate through the vein of the graft
B) An increase in venous pressure
C) Development of a stenosis at the arterial or venous end of the access
D) Placement of the arterial needle, near and directed toward, the anastomosis

63. Correct needle placement is important in the initiation of dialysis to minimize the risk of all of the following EXCEPT:
A) Recirculation
B) Steal syndrome
C) Inadequate blood flow
D) Hematoma formation

64. The dialysis procedure is terminated. Before heparinizing and capping of the catheter, the practioner should:
   A) Confirm placement of catheter with chest x-ray
   B) Follow the procedure for IV catheter care
   C) Check manufacturer's specifications for catheter lumen volume
   D) Change the catheter exit site dressing.

65. The most common complication of an AV fistula is:
   A) Infection
   B) Thrombosis
   C) Radial artery steal
   D) Aneurysm

66. The primary purpose of instilling a thrombolytic medication into a catheter is to:
   A) Prevent a myocardial infarction
   B) Give antibiotics access to bacteria
   C) Eliminate infection
   D) Improve the flow of solution

67. The percent recirculation for a patient with a systemic BUN of 85mg/dL, an arterial BUN of 70mg/dL and a venous BUN of 25 mg/dL is:
   A) 15%
   B) 20%
   C) 25%
   D) 30%

68. Which of the following statements regarding assessment of the vascular access is correct?
   A) Changes in the bruit on auscultation may indicate stenosis
   B) Decreased venous pressure is a common normal finding
   C) Decreased percent recirculation may indicate stenosis
   D) Increased cannulation difficulty is a common, normal finding

69. In explaining recirculation studies to a new staff member, which of the following would you state?
   A) The acceptable percent recirculation is less than or equal to 25
   B) The acceptable percent recirculation is less than or equal to 35
   C) The non-invasive method of performing recirculation studies may be more accurate than the traditional method
   D) The non-invasive method of performing recirculation studies is not as accurate as the traditional method

70. A graft should be cannulated, by inserting the needle at an angle between:
A) 10-15 degrees
B) 20-30 degrees
C) 35-45 degrees
D) 50-60 degrees

71. Mr. Jones has a left lower arm AV loop graft that has functioned well for 2 ½ years. Recently, there has been a concern that the percent recirculation in the graft may be increasing. Your concern is based on your observation of:
A) The color of the distal extremity
B) Unexplained increases in BUN & creatinine
C) Increased heparin requirements
D) Frequent clotting in the dialyzer

72. Mr. Jones' percent recirculation is high. An angiogram is performed. Restenosis is noted in the venous end of the graft. In an attempt to prevent as much recirculation as possible, you place the venous needle:
A) With the flow
B) Before the stenosis
C) Against the flow
D) After the stenosis

73. Mr. Jones has angioplasty performed and the percent recirculation in the access decreases. To detect future problems, you pay most attention to:
A) Arterial pressures
B) Dialysate pressures
C) Venous pressures
D) Transmembrane pressures

74. Ms. Grace has an AV graft placed in her right upper arm. She complains of pain in her right hand that increases in intensity during hemodialysis. To decrease the pain you would:
A) Keep the hand warm & dependent
B) Keep the hand warm & elevated
C) Keep the arm extended & elevated
D) Keep the arm extended & dependent

75. Grafts generally should NOT be cannulated for at least _____ weeks after placement and not until swelling has subsided so that palpation of the graft can be performed.
A) 2 weeks
B) 4 weeks
C) 6 weeks
D) 8 weeks

76. Which of the following is a risk factor for vascular access complications?
A) Presence of an AV fistula
B) Serum albumin of 3.8g/dL
C) Rotation of cannulation sites at each treatment
D) History of hypotensive episodes

77. Which procedure is necessary to be performed after the placement of a temporary or permanent catheter?
A) Chest x-ray
B) CAT scan
C) IVP
D) KUB

78. Mr. D's internal jugular catheter works well for the first two hours of his treatment but suddenly does not provide adequate blood flow. The practitioner's first action, aside from checking the patient's blood pressure, would be to:
A) Change the patient's dressing
B) Irrigate the catheter
C) Change the patient's position
D) Reposition the catheter

79. Mr. S's AV fistula has developed well and has now been used successfully for one week. The practitioner's concern for access longevity is best reflected in which of the following behaviors?
A) Attention to venous resistance
B) Rotation of cannulation sites
C) Manipulation of venous resistance
D) Distance between cannulation sites

80. Non-thrombotic catheter occlusions are caused by:
A) "Pinch-off Syndrome"
B) Catheter malposition
C) Precipitates from IDPN, lipids or medications
D) All of the above

81. The main complication(s) of Central Venous Access Devices is (are):
A) Mechanical occlusion
B) Catheter related infection
C) Thrombotic occlusion
D) All of the above

82. The primary sign or symptom of central vein stenosis is:
A) Hematoma formation
B) Warmth & hardness at the insertion site
C) Gross edema of the entire arm
D) Numbness & tingling of the distal fingers

83. The three (3) components of the access physical exam are:
A) Inspection, palpation and manipulation
B) Occlusion, auscultation and percussion
C) Inspection, palpation and auscultation
D) Manipulation, palpation and percussion

84. All grafts have a higher incidence of thrombosis then an AV fistula.
A) True
B) False

85. A properly functioning AV graft should have:
A) A continuous thrill at the arterial anastomosis, a soft compressible pulse and a low pitched continuous bruit
B) A continuous thrill throughout the graft, abounding pulse and a low pitched discontinuous bruit
C) A continuous thrill at the arterial anastomosis, a soft compressible pulse and a high pitched continuous bruit
D) A continuous thrill at the venous anastomosis, abounding pulse and a low pitched discontinuous bruit

86. Superficial infections of AV grafts are generally associated with cannulation.
A) True
B) False

87. Deep infections of AV grafts usually require surgical intervention.
A) True
B) False

88. Mr. Smith has had angioplasty on his left lower arm graft. The venous pressure and percent recirculation in the access is within normal limits. Inhibition of clothing becomes a concern, therefore, the practitioner pays particular attention to the prevention of:
A) Hypertension
B) Hypercalcemia
C) Hypotension
D) Hypocalcemia

89. Which of the following methods define recommended K-DOQI first-choice surveillance methods and frequency for optimal vascular (AVF or AVG) access assessment?
A) Color-flow Doppler once per quarter
B) Static Venous Pressure minimally every 2 weeks
C) Dynamic Venous Pressure every hemodialysis treatment
D) Dilution Technique once per quarter
E) On-line Clearance (OLC) once per quarter
F) All of the above EXCEPT C

Section VII
DIALYZER REPROCESSING

1. The concentration of the mixed Renalin used in the Renatron should be:
   A) 10%
   B) 33%
   C) 0.9%
   D) 21%

2. Dialyzer caps should be soaked in Renalin at a concentration of:
   A) 5%
   B) 1%
   C) 25%
   D) 21%

3. The following are risks and hazards of reprocessing.
   I. Sepsis, infection
   II. Pyrogenic reaction
   III. Reactions to reprocessing chemicals
   IV. Altered immune system
   V. Infusion of particulate
   VI. Increased dialyzer blood leak
   A) I and II only
   B) I, II, and III only
   C) I, II, IV, and VI only
   D) All of the above

4. The purpose of disconnecting the used dialyzer from the blood tubing and capping both blood ports is (are) to:
   A) Avoid introduction of bacteria
   B) Avoid introduction of air into the blood compartment
   C) Prevent contamination of the unit
   D) All of the above

5. All of the following solutions are commonly used to clean residual blood from the dialyzer <u>EXCEPT</u>:
   A) Hydrogen peroxide
   B) Bleach
   C) Peracetic acid
   D) Formaldehyde

6. Measurement of the total cell volume (TCV) is used to determine:
   A) Performance of the dialyzer
   B) Blood leak
   C) Contaminants
   D) Residual chemicals

7. A 20% decrease in total cell volume (TCV) results in a drop in dialyzer clearance of:
A) 5%
B) 10%
C) 20%
D) 25%

8. The total cell volume must be no less than:
A) 20%
B) 50%
C) 80%
D) 100%

9. The blood path integrity is tested to prevent the possibility of a blood leak. The drop in pressure should not exceed:
A) 5 mmHg
B) 10 mmHg
C) 20 mmHg
D) 25 mmHg

10. A positive or negative change in the ultrafiltration rate (K_{UF}) of the dialyzer should not exceed:
A) 10%
B) 20%
C) 25%
D) 50%

11. Any dialyzer that does not pass the TCV/leak test is a failed dialyzer and must be discarded.
A) True
B) False

12. Reprocessing records are considered medical records and should be maintained in a complete, legible, and secure manner.
A) True
B) False

13. Erythropoietin is known to reduce the effective clearances of reused dialyzers as a result of an increase in the hematocrit.
A) True
B) False

14. In order to avoid cross contamination in Hepatitis B antigen positive patients, reprocessing of their dialyzers should not be performed.
A) True
B) False

15. Which of the following actions should be taken if a wrong dialyzer is used for a patient?

    I. The patient's virology samples should be drawn
    II. Discontinue dialysis immediately
    III. The staff who put the patient on dialysis should be interviewed

A) I and II only
B) II and III only
C) I and III only
D) All of the above

16. Reprocessing polysulfone dialyzers with bleach will result in:

    I. A steady increase in β2 microglobulin removal over the usage
    II. Complement activation and leukopenia occur as with the first use
    III. An increase in albumin loss

A) I and II only
B) II and III only
C) I and III only
D) All of the above

17. The main purpose of completely disinfecting the O ring and header of a disassembled dialyzer before it is reassembled is:

A) To prevent bacterial growth where the disinfectant does not penetrate
B) To lubricate the O ring
C) To soften the O ring for assembly
D) All of the above

18. If dialysis is not commenced immediately following an acceptable residual germicide test, the following will occur EXCEPT:

A) Concentrations of germicide in the blood compartment increase
B) Dialyzer clearance diminishes
C) Re-equilibrium occurs between the potting compound and the saline in the blood compartment
D) The patient may exhibit symptoms of acute toxicity when dialysis is started

19. Germicide rebound is avoided by:

    I. Adequate rinsing of the dialyzer
    II. Performance of a residual germicide test
    III. Delaying dialysis for one hour
    IV. Discarding the priming saline when initiating dialysis

A) I, II, and III only
B) II, III, and IV only
C) I, II, and IV only
D) All of the above

20. Endotoxins are not only released when bacteria are killed; living bacteria also release endotoxins during reproduction and growth.
A) True
B) False

21. Pyrogens are any agent that cause fever when injected into humans.
A) True
B) False

22. Salmonella typhosa needs to reach 1 ng/kg before crossing the threshold and Escherichia coli only needs about 0.1 ng/kg to cause a pyrogenic response.
A) True
B) False

23. The use of new membranes with greater permeability to large molecules (4000-5000 Daltons) never has any effect on the transfer of bacterial proteins and other noxious byproducts in the dialysate to the patient.
A) True
B) False

24. Short dialysis time is associated with a greater post dialysis urea rebound.
A) True
B) False

Match the word to the correct statement

25. ___ germicide
26. ___ total cell volume
27. ___ automated reprocessing
28. ___ composite reimbursement
29. ___ cellulose

A) The fees paid by Medicare for each dialysis
B) A fiber that forms cell walls of many plants
C) A chemical that destroys bacteria
D) A machine to rinse, clean, and test dialyzers
E) A measure of how much fluid the dialyzer can hold on the blood side

30. To make sodium hypochlorite (Clorox Bleach 5.25%) 100 ppm, add 2.0 ml to 1000 cc of water. To make 10 ppm of bleach solution add ___ ml of bleach to 1000 cc of water.
A) 2.0
B) 0.2
C) 0.02
D) 0.002

31. To make 2% formalin (37% formaldehyde), add 54.0 mL in 1000 mL of water. To make 3% formalin solution add ___ mL of formaldehyde to 1000 mL of water.
A) 108.0
B) 7.6
C) 81.0
D) 10.8

32. An acid or vinegar rinse helps:
   A) Disinfect the hydraulic system
   B) Prevent precipitation
   C) Sterilizing the equipment
   D) All of the above

33. Dialyzer reprocessing requires:
   A) Water trace analysis
   B) Bacteria testing
   C) Endotoxin testing
   D) All of the above

34. The minimum standard for the level of disinfection to be used on a blood compartment and contacting surface is:
   A) Low level disinfection
   B) High level disinfection
   C) Sterilization
   D) None of the above

35. Clots are removed from the fibers during the DRS4 reprocessing procedure during the:
   A) Reverse ultrafiltration cycle
   B) Back filtration cycle
   C) Rinse cycle
   D) Performance tests

36. Your center reprocesses dialyzers. What are some possible clinical indications of decreasing efficiency that would alert the caregiver?
   A) An unusual rise in predialysis plasma urea or creatinine
   B) General deterioration of the patient's clinical condition
   C) An unanticipated deviation of the post dialysis weight, if a conventional fluid removal system was used
   D) All of the above

37. The number of times a dialyzer is reused is determined by:
   A) Reprocessing procedures
   B) Dialyzer priming procedures
   C) Proper heparinization
   D) All of the above

38. Patients have the right to refuse dialyzer reprocessing.
   A) True
   B) False

39. The term 510(K) is:
    A) Larger than a 401K
    B) An FDA approval for a medical device
    C) Not needed for hemodialysis equipment
    D) None of the above

40. The State Department of Health can:
    A) Recommend a dialysis facility follow a procedure
    B) Close the facility until they are compliant
    C) Start legal action
    D) All of the above

41. The steps, in order, for reprocessing dialyzers are:
    A) Test, rinse, clean, documentation, disinfect, statistical analysis
    B) Rinse, test, clean, disinfect, documentation, statistical analysis
    C) Documentation, clean, rinse, test, disinfect, statistical analysis
    D) Rinse, disinfect, clean, test, documentation, statistical analysis

42. Part of the rinse process during reprocessing is:
    A) Ultrafiltration
    B) Reverse ultrafiltration
    C) Blood side rinse
    D) B and C only

43. After the disinfect cycle, the dialyzer should be tested for:
    A) Residual disinfectant
    B) Total cell volume
    C) Leak test
    D) Presence of disinfectant

44. Before a patient is treated on a reprocessed dialyzer, the dialysis caregiver should perform a:
    A) Test for presence of disinfectant
    B) Leak test
    C) Test for residual disinfectant
    D) B and C only

45. Total cell volume equals:
    A) Fiber bundle volume
    B) Blood cap volume
    C) Dialysate compartment volume
    D) A and B

46. A reprocessed dialyzer can be used:
   A) On any patient
   B) On the same patient
   C) Once and discarded
   D) Up to one year

47. Water used for reprocessing dialyzers must meet:
   A) Local municipal water standards
   B) AAMI standards
   C) ANSI standards
   D) Facility preference

48. Verification of the test for residual disinfectant should be performed by:
   A) The practitioner initiating dialysis
   B) The patient
   C) The charge nurse
   D) Two individuals

49. The reuse technician is the only factor in the reuse average.
   A) True
   B) False

50. If the reuse average is low for one week in a clinic, the reuse technician should:
   A) Be dismissed
   B) Be retrained immediately
   C) Notify the charge nurse that there is a problem with the priming procedure
   D) Understand that this is a normal fluctuation of a process

51. Dialyzers that are primed and pass the test for residual may fail the test if dialysate flow and blood side recirculation is not maintained.
   A) True
   B) False

52. Germicide rebound is thought to occur because of:
   A) Residual germicide in the O ring
   B) Germicide bound to the dialyzer casing
   C) Inadequate dialyzer priming process
   D) Leaching from the potting compound

53. K-DOQI recommends measuring the TCV of every reuse dialyzer prior to its' first use, for all of the following reasons EXCEPT:
A) TCV can vary by 8% to 10% from the TCV stated by the manufacturer
B) There is lot to lot variability and hemodialyzer to hemodialyzer variability within a single lot
C) Measurement of TCV prior to hemodialyzer use provides a baseline value for subsequent monitoring and evaluation of the performance of each individual dialyzer
D) TCV measurement will predict the blood leak of the dialyzer

54. The OSHA Environmental Exposure Limits for glutaraldehyde are:
A) 0.2 ppm
B) 10 ppm
C) 1 ppm
D) 2 ppm

55. A toxic substance produced and held with the cell walls of bacteria until they die or are destroyed, whereupon it may be released.
A) Pseudomonas
B) LAL
C) E. Coli
D) Endotoxin

56. A measure of how much fluid a dialyzer can hold on the blood side is the:
A) Total cell volume
B) Dialyzer clearance
C) Ultrafiltration rate
D) Residual test

57. The OSHA Environmental Exposure Limits for formaldehyde are:
A) 0.2 ppm
B) 5 ppm TWA
C) 0.75 ppm TWA and 0.5ppm action level
D) 1 ppm TWA

58. With regard to evaluating the reusability of a hollow fiber dialyzer, which one of the following statements is correct?
A) Cell volume correlates with surface area which in turn correlates with dialyzer effectiveness
B) Compliance correlates with cell membrane volume which in turn correlates with patient symptoms
C) Surface area correlates with ultrafiltration which in turn correlates with re-usability
D) Dialyzer effectiveness correlates with priming volume which in turn correlates with patient symptoms

59. According to federal law, the reused dialyzer MUST be labeled with the:
A) Patient's name, the number of previous uses and the date of the last reprocessing
B) Patient's name, the ultrafiltration co-efficient and the patient's social security number
C) Patient's name, social security number and diagnosis
D) Patient's social security number, pressure test results and blood type

60. Careful rinsing of a re-used dialyzer prior to its being used for a patient is done primarily to remove:
A) Residual disinfectant solution
B) Residual blood
C) Bacteria
D) Cellular deposits

61. Dialyzer reuse is effected by:
A) Reprocessing procedures
B) Dialyzer priming procedures
C) Heparinization
D) All of the above

Section VIII
WATER TREATMENT

1. The main group(s) of water contaminants is (are):
   A) Microbiological
   B) Inorganic
   C) Organic
   D) All of the above

2. Water can be considered safe for drinking, but not necessarily safe for dialysis treatments because the:
   A) Patient exposure to water is 25 times greater
   B) Patient's urine output is decreased
   C) Fluid removal is a problem in renal patients
   D) All of the above

3. Aluminum, fluoride, and chloramine chemicals are added by municipalities to drinking water supplies.
   A) True
   B) False

4. The following are reasons for water treatment in dialysis units:
   - I. Impurities and toxins can foul dialysis machines
   - II. Patients are exposed to a large amount of water
   - III. Impurities may enter the blood through the dialyzer and can cause toxicity
   - IV. Calcium and magnesium can cause a hard water syndrome

   A) I and II only
   B) II and III only
   C) I, II, and IV only
   D) All of the above

5. Chloramine exposure during hemodialysis may cause:
   A) Hemolysis
   B) Hypernatremia
   C) Pericarditis
   D) Bleeding

6. In order to find the volume of supply water having a total hardness of 10 grains/gallon (as $CaCo_3$), which can be softened by a softener having the capacity of 100,000 grains (Assume 70% of maximum soften capacity), the following method is used:
100,000 grains x .70=70,000 grains
70,000 grains/10 grains/gal = 7000 gallons Answer: 7000 gallons

Using the information above, determine the volume of water having a total hardness of 120 mg/L (or 7 grains/gal*), which can be softened by a softener having the capacity of 15,000 grains (Assume 70% of maximum soften capacity)

* 1 grain/gal=17.1mg/L

A) 840 gallons
B) 1500 gallons
C) 2502 gallons
D) 2945 gallons

7. Softening is defined as the exchange of calcium and magnesium ions for sodium ions.
A) True
B) False

8. During softening, sodium levels are increased in proportion to the amount of calcium and magnesium removed.
A) True
B) False

9. According to AAMI standards, the total microbial count of water used to prepare dialysate:
A) Shall not exceed 200 cfu/mL
B) Shall not exceed 400 cfu/mL
C) Shall not exceed 1000 cfu/mL
D) Shall not exceed 2000 cfu/mL

10. According to AAMI standards, the total microbial count for dialysate:
A) Shall not exceed 1000 cfu/mL
B) Shall not exceed 2000 cfu/mL
C) Shall not exceed 200 cfu/mL
D) Shall not exceed 400 cfu/mL

11. According to AAMI standards, bacteriological testing for water and dialysate should be done:
A) Daily
B) Weekly
C) Monthly
D) Yearly

12. According to AAMI standards, when water purification devices are used, testing for chemical contaminants in the water should be done at least:
    A) Weekly
    B) Monthly
    C) Quarterly
    D) Yearly

13. Testing for chloramine levels in a dialysis unit is done by:
    A) Measuring for free chlorine
    B) Adding free chlorine and total chlorine
    C) Subtracting free chloramine from total chlorine
    D) Measuring for total chloramine

14. Chloroform and reactive agents, such as hypochloride, are the main contaminants removed by carbon absorption.
    A) True
    B) False

15. According to the Association for Advancement of Medical Instrumentation (AAMI), to remove chloramines effectively, the minimum empty bed contact time (EBCT) in the carbon tank is at least:
    A) 6 minutes
    B) 10 minutes
    C) 12 minutes
    D) 15 minutes

Select the correct statement from below to match:
(A) Cellulose membrane (B) Polyamide membrane (C) Thin film composites membrane.

16. ___Bleach can be used to disinfect the system
17. ___Tolerance of a wide pH range
18. ___Subject to degradation by oxidants (such as free chlorine)

19. As the total dissolved solids (TDS) contents of the water increases, the product water recovery:
    A) Decreases
    B) Increases
    C) Remains the same
    D) Increases over time

20. Bacterial exposure from water may cause which of the following symptoms?
    A) Hemolysis
    B) Fever, chills
    C) Bone disease
    D) Hypotension

21. The following can be removed by filtration:
A) Sodium
B) Chloramine
C) Chlorine
D) Bacteria

22. Carbon tanks work on the principle of:
A) Backwashing
B) Osmosis
C) Adsorption
D) Ion exchange

Match the following components with monitoring of the reverse osmosis system.

23.	___ Storage tank	A) LAL (Limulus amebocyte lysate)
24.	___ Deionizer	B) Resistivity
25.	___ Carbon	C) Hardness
26.	___ Reverse osmosis	D) Bacteria
27.	___ Reuse water	E) Chlorine
28.	___ Softener	F) Percent rejection

29. The filter that removes particulates from the water is called:
A) Backfiltration
B) Multimedia filtration
C) Ultrafiltration
D) Softener

30. Sediment filtration should have an opaque housing:
A) To inhibit proliferation of algae
B) To protect the filter from light
C) For aesthetic purposes
D) To keep it cool

31. A minimum flow velocity of 1.5 ft/sec through the feed line is needed to:
A) Reach all stations in the dialysis unit
B) Maintain pressure in the RO system
C) Discourage bacterial colonization
D) None of the above

32. The correct frequency and method of monitoring exhaust levels for DI tanks are:
A) Daily with audio alarms
B) Daily with audio and visual alarms
C) Weekly with visual alarms
D) Weekly with audio and visual alarms

33. The minimum required resistivity measurement for a DI tank is:
A) 24 D.C. volt
B) 110 volt
C) 1 megohm-cm
D) None of the above

34. A mixed-bed deionizer tank contains:
A) Cations
B) Anions
C) Activated carbon
D) Cations and anions

35. One megohm is equivalent to an ionic concentration of:
A) 0.500 ppm
B) 13.4 mS/cm
C) 1 ppm
D) 14.0 mS/cm

36. The minimum amount of endotoxin that will cause a LAL preparation to clot is called the LAL:
A) Sensitivity
B) Endotoxin
C) Culture
D) Cfu/ml

37. If the LAL has a sensitivity of 0.25 EU/mL and the unknown sample contains 0.5 EU/mL; a test of the undiluted sample would be positive. If the diluted sample is diluted four fold so that it now contains 1/4 the original endotoxin, the test for the 1:4 dilution result would be:
A) Negative
B) Positive
C) 1 CFU/mL
D) 5 EU/mL

38. The water sample has a conductance of 25 microSiemens/cm. What is the resistivity of that sample in megohms-cm?
A) 0.004
B) 0.2
C) 0.04
D) 4.0

39. The water sample has a conductance of 1200 microSiemens/cm. What is the conductance in ppm?
A) 450
B) 1800
C) 900
D) 720

40. The chemical restoration of ion exchange resins by treating them with acid and caustic is known as:
A) Backwashing
B) Rebuilding
C) Regeneration
D) Ion exchange

41. Soft water generally contains less than:
A) 2.0 grains/gallon of hardness
B) 2.5 grains/gallon of hardness
C) 1.0 grains/gallon of hardness
D) 17.1 grains/gallon of hardness

42. A reverse osmosis system has a feed water conductivity reading of 300 microSeimens/cm and a permeate conductivity reading of 15 microSiemens/cm. What is the percent salt rejection?
A) 85
B) 98
C) 45
D) 95

For questions 43-46, indicate which of the following statements are TRUE or FALSE concerning the flow of water in a pipe.

43. Water moves the fastest in the middle of the pipe.
A) True
B) False

44. Laminar flow is faster than turbulent flow
A) True
B) False

45. Pressure drops continuously along a pipe in the direction of fluid movement.
A) True
B) False

46. If the diameter of a pipe is doubled, the amount of flow it will allow also doubles.
A) True
B) False

47. If a pyrogenic reaction is suspected, the water should be cultured for:
A) Gram negative bacteria
B) B2 microglobulin
C) Gram positive bacteria
D) Limulas amebocyte lysate

48. Granular Activated Carbon is acid washed to accomplish all of the following <u>EXCEPT</u>:
A) To remove the ash
B) To increase the porosity
C) To remove the glucose
D) To increase the adsorbency

49. During the process of water softening:
A) Cations are exchanged for hydrogen ions and anions for hydroxyl ions
B) Calcium and magnesium ions are exchanged for sodium ions
C) Chloramine is exchanged for calcium ions
D) Solvents are removed for a mixture of solutes and solvents

50. If the level of aluminum in water exceeds the maximum allowable level, the patient may experience:
A) Fever and chills
B) Encephalopathy
C) Anemia
D) Hard water syndrome

51. If the level of copper in water exceeds the maximum allowable level, the patient may experience:
A) Anemia
B) Encephalopathy
C) Neuropathy
D) Osteomalacia

52. If the level of lead in water exceeds the maximum allowable level, the patient may experience:
A) Anemia
B) Osteomalacia
C) Neurological disease
D) Fever and chills

53. To test for residual chlorine in a home hemodialysis water treatment system, the patient is taught to collect the water sample after it has passed through the:
A) Carbon filter
B) Reverse osmosis
C) Deionizer
D) Sediment filter

54. In a water purification process, reverse osmosis is accomplished by:
A) Ultrafiltration under high pressure against a concentration gradient
B) Simultaneous cation and anion exchangers
C) Forcing water through a 5 micron prefilter
D) A cation exchanger that exchanges sodium for calcium and magnesium

55. A flushed sensation, headache, profound thirst and vomiting are all symptoms indicating that the dialysate water is high in:
A) Sodium
B) Aluminum
C) Zinc
D) Calcium

56. In dialysis, the continuous use of untreated water with high concentrations of calcium causes which one of the following problems?
A) Hyperparathyroidism
B) Hypothermia
C) Hypocalcemia
D) Hard water syndrome

57. What are the toxic effects or symptoms of chloramines in dialysis water?
A) Pyrogen reaction and lowered patient temperature
B) Disseminated intravascular coagulation (DIC) and sickle cell anemia
C) Weakness, lethargy, and seizures
D) Hemolysis, anemia, and methemoglobinemia

58. Chloramines are <u>MOST</u> effectively removed by which one of the following components of a water treatment system?
A) Ionization
B) Reverse osmosis
C) Activated carbon
D) Ultrafiltration

59. According to AAMI Standards and the American National Standards Institute (ANSI), the manufacturer of the water treatment system for dialysis is responsible for:
A) Ensuring that the water produced by the system is below the maximum allowable chemical contaminant levels at the time of installation
B) Maintaining the safety records for the facility's water
C) Ensuring a continuous acceptable water-purity level
D) Contacting state and/or local authorities about the facility's water

60. When a deionizer system is used to prepare water for dialysate, it should be monitored continuously to produce water of:
A) 1 MegOhm-cm or greater specific resistivity at 25° C
B) 0.1 MegOhm-cm or greater specific resistivity at 25° C
C) 0.5 Meg-Ohm-cm or greater specific resistivity at 25° C
D) 5 MegOhm-cm or greater specific resistivity at 25° C

61. Which one of the following elements, when present at excessive levels in water used in dialysis, can induce nausea, vomiting, hypertension and muscular weakness in the dialysis patient?
A) Chloride
B) Copper
C) Calcium
D) Cadmium

62. Mechanical filtration of suspended particulate matter includes all of the following EXCEPT:
A) Carbon filters
B) Membrane filters
C) Cartridge filters
D) Sand and gravel filters

63. A system of water treatment that removes pyrogens as well as solutes is called:
A) Reverse osmosis
B) Sediment filtration
C) Softening
D) Deionization

64. In the water treatment system, a semipermeable membrane and pressure process that removes both organic matter and electrolytes from processed water is:
A) Softening
B) Reverse osmosis
C) Deionization
D) Filtration

65. The primary purpose of carbon prefiltration of water is:
A) Removal of bacteria prior to the reverse osmosis process
B) Removal of free sodium absorbed by the water during the softening process
C) Neutralization of noxious odors
D) Removal of chlorine and chloramine

66. Shortly after the Monday morning dialysis shift begins, three patients on one side of the unit develop fever, chills, and malaise. The nurse in charge suspects bacterial contamination of the water treatment system and immediately pages the physician on call. The technician should:
   A) Administer two Tylenol tablets and a bolus of normal saline to each affected patient
   B) Put all machines in bypass, draw blood cultures and take samples for cultures from the dialysate and the water
   C) Return the patients' blood and send blood specimens to the laboratory for CBC and differential
   D) Discontinue dialysis by discarding the blood lines and dialyzers

67. Many municipal water treatment facilities reduce bacterial levels through addition of what chemical to treated water?
   A) Chloramine
   B) Formaldehyde
   C) Renalin
   D) Fluoride

68. During the process of deionization:
   A) Cations are exchanged for hydrogen ions and anions for hydroxyl ions
   B) Calcium and magnesium ions are exchanged for sodium ions
   C) Chloramine is exchanged for calcium ions
   D) Solvents are removed for a mixture of solutes and solvents

69. During reverse osmosis:
   A) Cations are exchanged for hydrogen ions and anions for hydroxyl ions
   B) Calcium and magnesium ions are exchanged for sodium ions
   C) Chloramine is exchanged for calcium ions
   D) Solvents are removed from a mixture of solutes and solvents

70. The water distribution system in a dialysis unit keeps bacterial levels low by:
   A) Proper system design and installation
   B) Periodic maintenance and disinfection
   C) Maintaining a minimum flow velocity of 1.5 ft/sec
   D) All of the above

Select the correct statements below to match:

A) Softening
B) Deioninization
C) Reverse Osmosis
D) None of the above

71. _____ Exchange cations for hydrogen ions and anion for hydroxyl ions.
72. _____ Exchange calcium and magnesium ions for sodium ions.
73. _____ Involves removal of solutes by filtration.
74. _____ Exchange chloramines for calcium ions.

75. An LAL test is:
A) An assay for lysate
B) An assay for endotoxin
C) An assay for limulus
D) An assay for amebas

Match the contaminants to the clinical symptoms it may cause:

76. ___ Aluminum — A) Liver
77. ___ Iron — B) Encephalopathy
78. ___ Lead — C) Anemia
79. ___ Copper — D) Neurological disease
80. ___ Fluoride — E) Hard water syndrome
81. ___ Calcium/Magnesium — F) Osteomalacia
82. ___ Endotoxin — G) Fever/chills

83. Many municipal water treatment facilities reduce bacterial levels through addition of what chemical to treated water?
A) Formaldehyde
B) Renalin
C) Fluoride
D) Chloramines

84. What are the toxic effects or symptoms of chloramines in dialysis water?
A) Pyrogen reaction and lowered patient temperature
B) Disseminated intravascular coagulation (DIC) and sickle cell anemia
C) Weakness, lethargy and seizures
D) Hemolysis, anemia and methemoglobinemia

85. Testing for chloramines levels in a dialysis unit is done by:
A) Measuring for free chlorine
B) Adding free chlorine and total chlorine
C) Subtracting free chloramines from total chlorine
D) Measuring for total chlorine

86. Chloramine exposure during hemodialysis may cause:
A) Hemolysis
B) Hypernatremia
C) Pericarditis
D) Bleeding

87. Testing the effluent for chlorine/chloramines breakthrough should be done:
A) Before every patient
B) Daily
C) Twice a day
D) Monthly

88. The difference between "free chlorine" and "total chlorine" is the:
A) Iodine content
B) Ammonia content
C) Chloramines content
D) Chlorine content

89. In order to remove chlorine and chloramine effectively, the recommended "iodine number" of a GAC (granular activated carbon) should be at least:
A) 1000
B) 200
C) 500
D) 900

90. The AAMI standards have established a maximum allowable level for free chlorine at:
A) 0.5 mg/L
B) 1.0 mg/L
C) 5.0 mg/L
D) 0.1 mg/L

91. The AAMI standards have established a maximum allowable level for chloramines at:
A) 1.0 mg/L
B) 2.0 mg/L
C) 0.1 mg/L
D) 0.2 mg/L

92. When chlorine combines with ammonia from decomposing vegetation, it forms:
A) Algae
B) Chloramines
C) Moss
D) Yeast

93. The new action level for endotoxin concentration in the product water is:

A) 0.125 EU/ml
B) 1.0 EU/ml
C) 5.0 EU/ml
D) 10 EU/ml

Section IX
BIOMEDICAL/CHEMISTRY

1. Gamma rays are emanations that have:
   A) Mass but no charge
   B) Charge but no mass
   C) Neither mass or charge
   D) Both mass and charge

2. The kind of energy stored in a chemical bond is:
   A) Potential energy
   B) Kinetic energy
   C) Activation energy
   D) Ionization

3. If an acidic solution is added to a basic solution, the pH of the basic solution:
   A) Decreases
   B) Increases
   C) Remains the same
   D) Neutralizes

4. Which of the following are classified as chemical substances?
   A) Mixtures and solutions
   B) Compounds and solutions
   C) Elements and mixtures
   D) Compounds and elements

Match the words to their definitions or characteristics.

5. ___ Atom
6. ___ Molecules
7. ___ Element
8. ___ Compound
9. ___ Valence
10. ___ Molarity
11. ___ Osmole
12. ___ Buffers
13. ___ Cations
14. ___ Anions
15. ___ Water rejecting
16. ___ Water accepting

A) Physically present particles
B) Smallest particle obtainable by physical means
C) More than one type of atom present
D) Contain only one type of atom
E) Number of electrons gained, lost, or shared in a bond
F) Hydrophobic (i.e. alcohol)
G) Solutions that resist a change in pH
H) Unit of osmotic pressure
I) A negatively charged ion
J) A positively charged ion
K) Hydrophilic (i.e. sodium chloride)
L) The concentration of a solution, measured as the number of moles of solute per liter of solution

Match the electrolytes to their correct functions.

17. ___ Na^+
18. ___ Cl^-
19. ___ Ca^{++}
20. ___ Mg^{++}
21. ___ HCO_3^-
22. ___ Acetate
23. ___ Glucose
24. ___ K^+

A) Major extracellular anion, osmotic pressure, pH balance
B) Major extracellular cation, osmotic pressure
C) Enzyme cofactor, intracellular cation
D) Muscle contraction, blood clotting, bone metabolism
E) Replacement buffer, converted in liver
F) pH balance, blood buffer
G) Cardiac rhythm, muscle contraction, major intracellular cation
H) Energy, fuel for nervous tissue

Match the following to an equivalent measure.

25. ___ 1 gallon
26. ___ 1 kg
27. ___ 1 mL
28. ___ 1 meter
29. ___ 1 liter
30. ___ 1 fluid ounce
31. ___ 37°C
32. ___ 100° F
33. ___ 1 lb
34. ___ 1 yard

A) 2.2 lbs
B) 3.78 L
C) 1000 cc
D) 36 inches
E) 98.6° F
F) 30 mL
G) 37.8° C
H) 39 inches
I) 1 cc
J) 0.454 kg

35. Calculate the approximate conductivity of the following solution: Sodium 137mEq/L, Calcium 3.5 mEq/L, Potassium 2.0 mEq/L, Magnesium 1.0 mEq/L
A) 13.70 mS/cm
B) 13.75 mS/cm
C) 13.50 mS/cm
D) 13.00 mS/cm

Match the word to the correct statement.

36. ___ Volt
37. ___ A.C.
38. ___ Ampere
39. ___ D.C.
40. ___ Impedance
41. ___ Resistance
42. ___ GFD

A) The unit of measurement for electrical current
B) Direct current
C) The type of current usually supplied by utility power lines
D) The unit of measurement of electromotive force
E) A monitor for any current between either power wire and ground
F) The total opposition a circuit offers to the flow of alternating current
G) A property of a conductor which causes it to impede the flow of electrons

43. The reference resistance used when referring to or measuring leakage current on medical equipment is:
A) 10 ohms
B) 100 ohms
C) 1000 ohms
D) 10,000 ohms

44. Normal leakage current on a dialysis machine should be less than:
A) 50 microamp
B) 100 microamp
C) 300 microamp
D) 500 microamp

45. The binary number of 10 is:
A) 1001
B) 1100
C) 1011
D) 1010

46. UL-1262 specifies that the maximum leakage current of a laboratory instrument should be:
A) 100 uA
B) 500 uA
C) 50 uA
D) 250 uA

Matching

47. ___ Binary
48. ___ Bit
49. ___ Byte
50. ___ CPU
51. ___ Disk
52. ___ Hexadecimal
53. ___ I/O
54. ___ Integrated circuit
55. ___ RAM
56. ___ ROM
57. ___ DOS

A) A magnetically coated substance on which data for a computer is stored
B) Central processing unit
C) The smallest quantity a computer can measure or detect
D) Random access memory
E) Disk operating system
F) The base two numbering system whose only digits are 0 and 1
G) Read only memory
H) The unit of measure for computer memory and disk storage
I) Input/output
J) The base sixteen numbering system whose digits go from 0 through F
K) Chip

58. Conductivity = Resistivity
    A) True
    B) False

59. Conductivity = 1/ Resistivity
    A) True
    B) False

60. Resistivity = 1/ Conductivity
    A) True
    B) False

61. V=IR
    A) True
    B) False

62. R = V/I
    A) True
    B) False

63. Pressure = Flow x Resistance
    A) True
    B) False

64. Venous pressure gauges are calibrated in:
  A) PSI
  B) mmHg
  C) mm/H20
  D) Mg%

65. TMP is measured in:
  A) mL/min
  B) mmHg
  C) mL/hour
  D) None of the above

66. Conductivity meters measure:
  A) Sodium content of dialysate
  B) Water content of dialysate
  C) Electrical flow through the solution
  D) Sodium and potassium content of the dialysate

67. Prior to dialysis, the dialysate should be tested for:
  A) Conductivity, pH, temperature
  B) Non-ionic content, pH, conductivity
  C) Temperature, glucose content, pH
  D) Glucose content, non-ionic content, pH

68. A milliequivalent measures the:
  A) Ion combining power
  B) Weight of the particle
  C) Molecular weight in milligrams
  D) Amount of charge of a solute

69. The pH is used to measure the concentration of:
  A) All ions in solution
  B) Hydrogen ions
  C) Bicarbonate ions
  D) Acetate ions

70. A solute, such as sodium chloride, when it ionizes in solution is called a (an):
  A) Ion
  B) Electrolyte
  C) Solvent
  D) Compound

71. A patient has a body temperature of 99.0 degrees F. Using the following formula, calculate the patients' temperature in degrees C.

Degrees C = (Degrees F - 32) x 5/9

A) 36.0° C
B) 37.0° C
C) 37.2° C
D) 37.7° C

72. The first action to take when a fire occurs is:
A) Sound the fire alarms
B) Extinguish the fire using appropriate extinguishers
C) Rescue those in the nearest vicinity
D) Confine the fire to one area

73. Some of the measures to take to prevent electrical shock are:
I. Never apply a 3-prong plug to a 2-prong adapter
II. Extension cords should not be used
III. Never pull on the power cord to remove a plug from its outlet
IV. Report equipment damage
A) I and II only
B) III and IV only
C) I, II, and IV only
D) All of the above

74. You witnessed a co-worker who sustained an electric shock from a piece of equipment, you should:
I. Unplug the equipment and call for help
II. Don't touch the victim
III. Remove the victim's hand from the equipment
IV. Turn off the equipment
A) I only
B) I and II only
C) II and IV only
D) All of the above

75. If the hospital goes on emergency power, you should:
I. Assure equipment is plugged into the red receptacle or emergency power outlet
II. Check equipment for proper function
III. Reset all alarms
IV. Reassure the patients
A) I and II only
B) III and IV only
C) I and IV only
D) All of the above

76. Three components of an ordinary fire are:
A) Smoke, flame, and heat
B) Fuel, heat, and oxygen
C) Gas, liquid, and vapor
D) Flame, sparks, and explosions

77. When a quantity of electricity is converted to heat, the heat energy is measured in:
A) Volts
B) Amperes
C) Calories
D) Degrees

78. Which of the following may be heterogeneous?
A) A substance
B) An element
C) A compound
D) A mixture

79. The atomic number of an atom is always equal to the total number of:
A) Neutrons in the nucleus
B) Protons in the nucleus
C) Neutrons plus protons in the atom
D) Protons plus electrons in the atom

80. A particle which is electrically neutral is a:
A) Proton
B) Ion
C) Neutron
D) Electron

81. Radioisotopes used for medical diagnosis should have:
A) Short half-lives and be quickly eliminated from the body
B) Short half-lives and be slowly eliminated from the body
C) Long half-lives and be quickly eliminated from the body
D) Long half-lives and be slowly eliminated from the body

82. When a chlorine atom reacts with a sodium atom to form a compound, the chlorine atom will:
A) Lose one electron
B) Lose two electron
C) Gain one electron
D) Gain two electrons

83. Which of the following elements has the strongest attraction for electrons?
A) Al
B) Cl
C) Si
D) Na

84. The attraction between water molecules and a Na+ ion or a Cl- ion occurs because water molecules are:
A) Linear
B) Symmetrical
C) Polar
D) Nonpolar

85. A pure substance melts at 38 degrees C and does not conduct electricity in either the solid or liquid phase. The substance is classified as:
A) Ionic
B) Metallic
C) Electrovalent
D) Molecular

86. A solution in which an equilibrium exists between dissolved and undissolved solutes must be:
A) Saturated
B) Unsaturated
C) Dilute
D) Concentrated

87. Which pH value represents a solution with the lowest OH- ion concentration?
A) 1
B) 7
C) 10
D) 14

88. Water containing dissolved electrolyte conducts electricity because the solution contains mobile:
A) Electrons
B) Molecules
C) Atoms
D) Ions

89. If A = BC/D, then:
A) B = AC/D
B) B = AD/C
C) B = DC/A
D) B = ACD

90. 300 mL/min divided by 15 mL/kg is equal to:
A) 20 mL/min/kg
B) 20 mL/kg
C) 20 kg/min
D) 20 kg/mL/min

91. Which of the following groups of compounds are listed from lowest pH to highest pH?
A) Acetic acid, R.O water, bicarbonate dialysate
B) Bleach, R.O. water, bicarbonate dialysate
C) Acetate dialysate, bicarbonate dialysate, R.O water
D) White vinegar, bleach, urine

92. If the specific gravity of a bicarbonate concentrate is 1.060, one liter of solution will weigh:
A) 1600 grams
B) 1000.6 grams
C) 1000.06 grams
D) 1060 grams

93. An acid with a pH of 2.0 will have a hydronium ion concentration 100 times greater than a solution with a pH of:
A) 4.0
B) 102.0
C) 0.0
D) 2.2

94. Standard temperature and pressure is:
A) 25° C and 760 mmHg
B) 25° C and 0 mmHg
C) 0° C and 760 mmHg
D) 0° C and 0 mmHg

95. 39.37 inches equals one meter, convert one foot into millimeters:
A) 30.48
B) 328.08
C) 32.8
D) 304.8

96. A micrometer is used to describe the size of a bacteria. How many micrometers does the average human hair measure?
A) 40 to 80 micrometers
B) 200 to 500 micrometers
C) 100 nanometers
D) 1 millimeter

97. A component which changes A.C. voltage into D.C. voltage is a:
A) Transformer
B) Bridge rectifier
C) Power transistor
D) Operational amplifier

98. If carbon weighs 12.011, hydrogen weighs 1.008, and oxygen weighs 16.00, how much does acetic acid (CH_3COOH) weigh?
A) 60.054
B) 58.038
C) 61.038
D) 84.076

99. If the size of a bubble in solution is directly proportional to temperature and inversely proportional to pressure, then which of the following statements is TRUE?
A) If dialysate solution temperature is increased, a bubble in the dialysate will shrink in size
B) As bubbles rise to the surface of the solution in an open container, its size will remain constant
C) Deaeration systems use positive pressure to create larger air bubbles
D) Bubbles in a dialyzer will become larger as the TMP is increased

100. If one volt is applied across two resistors, one of 10 ohms and one of 100 ohms, which of the following statements is FALSE?
A) The current through the 10 ohm resistor will be greater than through the 100 ohm resistor
B) The 100 ohm resistor will dissipate more heat
C) If the resistors were in series, the current through both would be the same
D) If the resistors were in series, the largest volt drop would be across the 100 ohm resistor

101. The median of the following numbers :17, 21, 27, 29, 32 is:
A) 25
B) 27
C) 25.2
D) 5

102. Burning of the eyes, lacrimation and general irritation to the upper respiratory passages are the first symptoms experienced when formaldehyde is inhaled in concentrations ranging from:
A) 0.1-5.0 parts per million
B) 5-10 parts per million
C) 10-20 parts per million
D) 20-50 parts per million

103. All of the following are necessary to ensure quality control in the area of electrical safety <u>EXCEPT</u>:
A) Monitoring for the correct electrical power entering the dialysis facility
B) Maintaining corrosion-free electrical equipment
C) Ensuring proper grounding of dialysis equipment
D) Periodic inspection of the condition of electrical cords and receptacles

104. An anion is:
A) A positively charged ion
B) Unit of osmotic pressure
C) A hydrophobic
D) A negatively charged ion

105. A cation is:
A) A negatively charged ion
B) A positively charged ion
C) Unit of osmotic pressure
D) A hydrophobic

106. The concentration of a solution measured as the number of moles of solute per liter of solution is:
A) Molarity
B) A compound
C) A molecule
D) A buffer

107. Solutions that resist a change in pH are:
A) Cations
B) Anions
C) Molecules
D) Buffers

108. An electrolyte that is a major extracellular anion that affects osmotic pressure and pH balance is:
A) Sodium
B) Chloride
C) Potassium
D) Bicarbonate

109. An electrolyte that is a major extracellular cation that affects osmotic pressure is:
A) Potassium
B) Sodium
C) Chloride
D) Calcium

110. An electrolyte that is a major intracellular cation and affects cardiac rhythm and muscle contraction is:
A) Potassium
B) Sodium
C) Calcium
D) Chloride

111. A cation that affects muscle contraction, blood clotting, and bone metabolism is:
A) Potassium
B) Sodium
C) Chloride
D) Calcium

112. The Association for the Advancement of Medical Instrumentation (AAMI) standards for blood leak detectors requires detection of less than 0.45mL/min of blood at hematocrit 45 over a range of dialysate flows.
A) True
B) False

Section X
LABS

1. To perform a creatinine clearance test requires:
   A) A timed measured urine collection and a serum creatinine measurement
   B) Three or more serum creatinine measurements and a single spot urine creatinine measurement
   C) Spot urine and serum creatinine measurement
   D) Urine measurements of protein, uric acid, and urea

2. A creatinine clearance provides information about the:
   A) Kidneys' ability to regulate renin production
   B) Kidneys' ability to regulate glomerular filtration rate
   C) Kidneys' ability to concentrate urine
   D) Kidneys' ability to reabsorb sodium in the Loop of Henle

3. Which one of the following laboratory findings is <u>NOT</u> affected by dietary intake and is used as an indicator of renal failure?
   A) Creatinine
   B) Sodium
   C) Blood urea nitrogen
   D) Potassium

4. A renal patient's serum potassium concentration should be maintained within a range of:
   A) 2.5-3.5 mEq/L
   B) 3.5-5.5 mEq/L
   C) 6.0-7.0 mEq/L
   D) 7.0-8.0 mEq/L

5. The acceptable range of serum albumin for dialysis patients is:
   A) 3.5-5.0 g/dl
   B) 2.5-3.5 g/dl
   C) 5.0-7.0 g/dl
   D) 1.5-3.0 g/dl

6. Elevated levels of this amino acid is associated with heart disease and a marker for left ventricular hypertrophy.
   A) Folic acid
   B) Homocysteine
   C) Transferrin
   D) Ferritin

7. The plasma protein responsible for iron exchange between tissues is:
   A) Iron
   B) Ferritin
   C) Transferrin
   D) Albumin

8. The best chemical indicators of iron availability are:
   A) Serum ferritin and serum transferrin saturation
   B) Iron and serum ferritin
   C) Iron and serum transferrin saturation
   D) Iron and total iron binding capacity

9. A measurement of the percentage of red blood cells in whole blood is the:
   A) Hemoglobin
   B) Hematocrit
   C) Transferrin Saturation
   D) Mean cell volume

10. Complement activation between the dialyzer membrane and blood that occurs during dialysis causes this lab test to be inaccurate if drawn during or after dialysis:
    A) Complete blood count
    B) Albumin
    C) Glucose
    D) Bicarbonate

11. The iron containing pigmented protein of a red blood cell is the:
    A) Hematocrit
    B) Hemoglobin
    C) White blood cell
    D) Ferritin

12. Hemoglobin is responsible for the:
    A) Transport of oxygen and carbon dioxide
    B) Coagulation of blood
    C) Indicator of available iron for erythropoeisis
    D) All of the above

13. The following lab test is used to monitor coumadin therapy:
    A) PT/INR
    B) PTT
    C) ACT
    D) Lee-White clotting time

14. A serum potassium is elevated when the level is above:
   A) 2.5 mEq/L
   B) 4.0 mEq/L
   C) 2.0 mEq/L
   D) 5.5 mEq/L

15. The normal serum pH range is between:
   A) 7.35-7.45
   B) 7.5-7.65
   C) 7.0-8.0
   D) 8.0-9.0

16. The normal serum bicarbonate level is:
   A) 10 mEq/L
   B) 20 mEq/L
   C) 24 mEq/L
   D) 32 mEq/L

17. Total cholesterol should be kept less than
   A) 100 mg/dL
   B) 300 mg/dL
   C) 50 mg/dL
   D) 200 mg/dL

18. The patient's plasma calcium level should be maintained between:
   A) 3.0-4.5mg/100 mL
   B) 4.5-6.0 mg/100 mL
   C) 8.5-9.6 mg/100 mL
   D) 13.0-14.0 mg/100 mL

19. The normal serum carbon dioxide content is:
   A) 4-10 mmHg
   B) 10-22 mmHg
   C) 36-48 mmHg
   D) 24-30 mmHg

20. Normal serum osmolality is:
   A) 101-200 mOsm/kg
   B) 285-295 mOsm/kg
   C) 325-415 mOsm/kg
   D) None of the above

21. The transferrin saturation level is normally:
  A) 10-25 mg
  B) 11-20 mg
  C) 55-100%
  D) 20-50%

22. The normal range of the anion gap is:
  A) 9-16 mEq/L
  B) 20-30 mEq/L
  C) 4-10 mEq/L
  D) None of the above

23. A patient's lab readings show a significant decrease in TSAT and TIBC levels with an increase in ferritin. His tests indicate:
  A) Inflammation
  B) Iron deficiency
  C) Erythropoiesis
  D) Iron overload

24. A hematocrit of 30% has been associated with:
  I Increased survival of dialysis patients
  II Decreased left ventricular hypertrophy
  III Improved quality of life
  IV Improved exercise capacity
  A) I and II only
  B) I, II, and III only
  C) II, III, and IV only
  D) All of the above

25. The following should be evaluated if a patient's hemoglobin/hematocrit is decreasing:
  A) Stool for guaiac
  B) Reticulocyte count
  C) Iron level
  D) All of the above

26. As renal function significantly decreases, laboratory data reflect all of the following EXCEPT:
  A) An increased BUN
  B) An increased hematocrit & red blood cell count
  C) An increased serum creatinine
  D) A decreased creatinine clearance

27. Which of the following laboratory markers suggests the potential for bone disease?

A) Hypercalcemia
B) Hypophosphatemia
C) Increased parathyroid hormone (PTH) level
D) Reduced PTH level

Section XI
MEDICATIONS/ANTICOAGULATION

1. In the clotting cascade, heparin interferes with:
   A) Platelet adhesion
   B) Platelet aggregation
   C) Conversion of prothrombin to thrombin
   D) Disintegration of clotting

2. Heparin is what type of medication?
   A) Anticoagulant
   B) Antihypertensive
   C) Antianemic
   D) Vitamin

3. After the initial bolus of heparin, the peak effect is reached:
   A) Immediately
   B) In approximately 20-30 minutes
   C) In approximately 5-10 minutes
   D) In an hour

4. The method by which heparin is given at intervals throughout the treatment is called:
   A) Continuous heparinization
   B) Intermittent heparinization
   C) Regional heparinization
   D) Hourly heparinization

5. The purpose of regional heparinization is:
   A) To systemically anticoagulate the patient
   B) To give only enough heparin to keep the dialyzer from clotting
   C) To anticoagulate just the blood in the extracorporeal circuit
   D) All of the above

6. Regional heparinization is performed by:
   A) The use of calcium citrate in the extracorporeal circuit
   B) Administration of heparin predialyzer, protamine post dialyzer
   C) Continuous heparin administration throughout treatment
   D) Administering heparin to the patient and citrate into the extracorporeal circuit

7. The amount of protamine sulfate that should be given to neutralize heparin is:
   A) 1-1.5 mg protamine/100 units Heparin
   B) 2 mg protamine/1000 units Heparin
   C) 1 unit protamine/1 units Heparin
   D) Dependent on patient weight

8. The expected activated clotting time (ACT) prior to heparinization is:
   A) 90-110 seconds
   B) 60 seconds
   C) 120-150 seconds
   D) Above 150 seconds

9. The expected change in the ACT after an initial dose of heparin is:
   A) 4x baseline
   B) 1.5-2x baseline
   C) 1 x baseline
   D) 3 x baseline

10. Heparin rebound is a potential complication of regional heparinization. This is caused by:
    A) An incorrect dose of protamine sulfate
    B) Improper utilization of heparin by the patient
    C) An allergic type reaction
    D) The RES (Reticulo-endothelial system) breaking down the heparin protamine complex

11. Patients with chronic renal failure should NOT take phosphate binders such as Amphogel or Basagel because they may cause:
    A) Constipation
    B) Bone decalcification
    C) Aluminum toxicity
    D) Diarrhea

12. Multivitamin supplementation is necessary for patients on hemodialysis because:
    A) Liver excretion is increased
    B) Water soluble vitamins are dialyzable
    C) Vitamins are good for them
    D) Vitamins are not absorbed in the GI tract

13. The most important factor to consider when a patient undergoing hemodialysis is on digoxin is:
    A) An additional dose of digoxin must be given post-dialysis to replace drug lost by dialysis
    B) The heparin given during dialysis may interact with digoxin and predispose to a bleeding episode
    C) A low potassium level may predispose the patient to digoxin toxicity
    D) Patient must take twice the normal amount of digoxin

14. Which of these antibiotics should be avoided in a patient with renal failure?
    A) Tetracycline
    B) Gentamycin
    C) Ampicillin
    D) Neomycin

15. A patient with chronic renal failure takes phosphate binders to:
    A) Prevent gastric irritation from drugs
    B) Prevent bone demineralization
    C) Increase serum phosphate
    D) Prevent constipation

16. The kidneys play a major role in the excretion of drugs from the body. Due to renal failure, which of the following statements is true concerning medications?
    A) The patient may need more than the usual therapeutic dose
    B) The patient may be at an increased risk of toxicity
    C) Certain medications may only require a third of the usual dose
    D) If a patient has a functioning liver, it will compensate for the renal failure

17. A basic principle to teach all chronic renal failure patients about medication is:
    A) Never take medications ordered by any physician other than the nephrologist
    B) Never take medications before dialysis
    C) Do not take over the counter medication without consulting the nephrologist
    D) Always take medications before dialysis

18. Aspirin can be dangerous for the renal failure patient if not monitored closely because it can cause:
    A) Metabolic acidosis
    B) Bleeding
    C) A change in mental status
    D) Clotting

19. The purpose of using Tums (calcium carbonate), Phoslo (calcium acetate), Fosrenol (lanthanum carbonate), Renagel (sevelamer hydrochloride) or Renvela (sevelamer carbonate) in patients with renal failure is to:
    A) Bind phosphorus
    B) Supplement calcium
    C) Decrease acidity
    D) All of the above

20. Multivitamins and folic acid are regularly prescribed because:
    A) Water soluble vitamins are lost during dialysis
    B) Patients usually do not get these vitamins in the foods they eat
    C) Patients are immunocompromised
    D) Patients with renal failure need three times as much of these medications

21. The primary reason for giving IV iron to hemodialysis patients with renal failure is:
A) Iron is lost during dialysis
B) There is not enough iron in the diet
C) Iron is necessary for erythropoiesis
D) Iron is needed for bone formation

22. To avoid toxicity from medications due to renal failure, what kind of adjustment may be made in the medication regimen?
A) Change in dosage and frequency
B) AM instead of PM
C) With meals instead of before meals
D) All of the above

23. A decrease in insulin requirements for diabetics on hemodialysis is typical due to a decreased:
A) Glucose intake
B) Tolerance for higher levels of glucose
C) Degradation of insulin
D) Half-life of insulin

24. Interventions used to maintain a patient's hemoglobin/hematocrit prior to the availability of erythropoietin was (were):
A) Transfusions
B) Androgens
C) Iron
D) A and B

25. The following medication should be avoided by a patient with renal failure:
A) Colace
B) Milk of magnesia
C) Dulcolax
D) Lactulose

26. PD patients taking iron supplements should be taught that it:
A) Is important for erythropoiesis
B) Should not be taken with phosphate binders
C) Should not be taken with milk
D) All of the above

27. A patient's phosphorus is 5.7. He is placed on phosphate binders. What instructions should he be given regarding phosphate binders?
A) Take only on an empty stomach

B) Take 1 hour after eating
C) Take with milk
D) Take with meals

28. Heparinization during hemodialysis can best be monitored by:
A) Bleeding
B) Clotting
C) Whole blood activated clotting time
D) Clotting in the dialysis circuit

29. The largest quantity of calcium in the body is located in the:
A) Extracellular fluid
B) Urine and stool
C) Intracellular fluid
D) Bones and teeth

30. In chronic renal failure, the following occurs:
A) As serum calcium increases, serum phosphate decreases
B) As serum calcium increases, serum phosphate increases
C) As serum calcium decreases, serum phosphorus increases
D) There is no change in serum calcium and serum phosphorus

31. The effect of calcitrol on phosphorus is to:
A) Reduce absorption from the gut
B) Increase absorption from the gut
C) Reduce absorption in the renal tubules
D) Increase absorption in the renal tubules

32. The two most common sources of aluminum contamination in hemodialysis patients are:
I. Calcium carbonate binders
II. Aluminum based binders
III. Inadequately treated water
IV. Antihypertensive medications
A) I and II only
B) II and III only
C) III and IV only
D) I and IV only

33. The risk of metastatic calcification increases when the calcium x phosphorus product exceeds:
A) 10.5-11.5
B) 75-80
C) 55
D) 3.5-4.5

34. The purpose of administering Calcijex IV for dialysis patients is to:
    A) Improve the calcium level
    B) Suppress the parathyroid gland
    C) Increase the Vitamin D level
    D) All of the above

35. An indication for stopping Calcijex administration is:
    A) Hypercalcemia
    B) Phosphatemia
    C) Hypocalcemia
    D) Anemia

36. A possible complication associated with the use of Alteplase is:
    A) Clotting
    B) Bleeding
    C) Hyperkalemia
    D) Hypercalcemia

37. Alternatives to systemic heparin anticoagulation include:
    A) Frequent saline rinses, tri-sodium citrate, and regional heparinization
    B) Intravenous dicumeral, tri-sodium citrate and calcium free dialysate
    C) Frequent saline rinses, intravenous vitamins and regional heparinization
    D) Heparinized priming solution, frequent saline rinses, blood flow rates less than 200 mls

38. Clotting in the dialyzer can be attributed to all of the following factors <u>EXCEPT</u>:
    A) Inadequate heparinization
    B) Inadequate blood flow rate
    C) Low hematocrit
    D) Low temperature of dialysate

39. Anticoagulation is necessary for patients undergoing hemodialysis for which of the following reasons?
    A) Blood clots quickly in the extracorporeal circuit
    B) Patients have low platelet counts
    C) Patients have hematocrits of 45%
    D) Patients are anemic

40. For a patient to be systemically heparinized, the loading dose should be given:
    A) Within 3-5 minutes of the start of the dialysis
    B) 3-5 minutes before treatment initiation
    C) When the treatment is calculated
    D) At the initiation of treatment

41. The rationale underlying the use of heparin during dialysis is:
   A) Anticoagulation is important to maintain oncotic pressure
   B) Blood coming in contact with foreign surfaces tends to initiate clot formation
   C) Blood coming in contact with foreign surfaces tends to increase surface tension
   D) Anticoagulation is important to improve solute drag

42. Kayexelate acts by:
   A) Forcing potassium from the intracellular space to the extracellular space
   B) Buffering acidosis
   C) Exchanging sodium ions for potassium ions in the GI tract
   D) Antagonizing the effects of potassium on the electrical conduction system of the heart

43. A patient is placed on Sensipar. Which of these might be a reason for Sensipar referral?
   A) The patient's phosphorus is high.
   B) The PTH is high and the calcium may run at a high level.
   C) Potassium is high.
   D) It is an appetite enhancer.

44. Minimal heparinization should be considered in patients who have a history of all of the following <u>EXCEPT</u>:
   A) Cataracts
   B) Menstrual flow
   C) Pericarditis
   D) Active ulcer

45. All of the following are relevant factors in ensuring the accuracy of the activated clotting time (ACT) <u>EXCEPT</u> the:
   A) Volume of blood in the ACT tube
   B) Patient's hematocrit value
   C) Temperature of the ACT tube
   D) Prompt admixing of blood and clotting reagent

46. Although the elimination of heparin in dialysis patients is similar to that of the non-uremic population (a half life of 40-90 minutes):
   A) Diabetic dialysis patients exhibit a heparin half-life up to three times longer than non-diabetic dialysis patients
   B) Platelet aggregation in the dialysis population is markedly increased, placing patients at risk for bleeding
   C) The "rebound effect" of heparin metabolism increases the incidence of bleeding in this population
   D) Hemodialysis patients may exhibit increased bleeding tendencies for up to 24 hours post dialysis

47. Phosphate binders should be taken:
A) Before or with meals
B) One hour after meals
C) Two hours after meals
D) Two hours before meals

48. Which of the following is among the reasons that oral iron is usually inadequate for administration to dialysis patients?
A) Oral iron seems to accumulate in the liver of dialysis patients
B) Due to increased RBC survival, oral iron cannot be absorbed by the cells and is eliminated through the stool
C) Dialysis patients usually take phosphate binders that bind iron as well as phosphates in the GI tract
D) Oral iron must be taken in conjunction with Vitamin D before adequate uptake will occur in the bowel

49. The strength of heparin most commonly used in hemodialysis is:
A) 10,000 units/mL
B) 5,000 units/mL
C) 1,000 units/mL
D) 100 units/mL

50. A physician orders a 2500 units loading dose of heparin and a 500 unit per hour infusion of heparin for 3.5 hours. How many total mLs of heparin (1000 units/mL) will be used during the course of treatment?
A) 0.425
B) 425
C) 4.25
D) 4250

51. In the previous question, how many units of heparin would be used during the course of treatment?
A) 0.425
B) 425
C) 4.25
D) 4250

52. A medication that prolongs the activity of heparin is:
A) Protamine sulfate
B) Aspirin
C) Iron
D) Folic acid

53. The practitioner must ensure that patients receiving digitalis therapy:
A) Dialyze against a dialysate potassium of at least 2.0 mEq/L
B) Dialyze against a dialysate potassium of 0
C) Do not take digitalis on dialysis days
D) Digitalis is increased on dialysis days

54. Which of the following anticoagulants is commonly used to reduce the risk of catheter-related thrombosis development?
A) Heparin
B) Enoxaparin
C) Deltaparin
D) Warfarin

55. The hemoglobin of a patient who has been successfully maintained on ESA therapy has decreased from 10.9g/dL to 9.9g/dL. The following should be done to determine what could be causing the decreased hematocrit.
A) Test for increased levels of folate and vitamin B_{12}
B) Test for decreased levels of aluminum
C) Test for increased levels of potassium
D) Test for decreased levels of iron

56. Which one of the following are possible causes of resistance to ESA therapy?
A) Aluminum overload and iron deficiency
B) Aluminum deficiency and iron overload
C) Folate deficiency and vitamin B_{12} toxicity
D) Vitamin B_{12} deficiency and folate toxicity

57. Currently, the most widely used agent for outpatient anticoagulation is:
A) Unfractionated heparin
B) Low molecular weight heparin (LMWH)
C) Warfarin
D) A Factor Xa inhibitor

58. For most patients taking coumadin, the target INR is:
A) A range of 1.5-2.0
B) 2.5 (range of 2.0 to 3.0)
C) 3.0 (range of 2.5 to 3.5)
D) 3.5 (range of 3.0 to 4.0)

59. Anticoagulants:
A) Break apart any existing clot
B) Prevent the development of new clots
C) Dissolve clots
D) Cannot halt enlargment of an existing clot

60. Which of the following would be indicative of excessive administration of heparin?
A) Prolonged bleeding
B) Low-blood pressure
C) Patient cramping
D) Clear dialyzer

61. Heparin is neutralized by protamine sulfate because:
A) Heparin is alkaline and protamine is acidic and they neutralize each other to form a salt
B) Protamine has a higher molecular weight than heparin
C) Heparin is acidic and protamine is alkaline and they neutralize each other to form a stable salt
D) Protamine is a polysaccharide and heparin is cholesterol

62. Heparin is utilized in catheters to prevent clotting and maintain the patency of the catheter lumen. Sometimes Alteplase (TPA) is used in the catheter. What is the action of TPA?
A) It prevents infections
B) It stops the formation of crystals
C) It helps degrade fibrin clots
D) It helps to maintain the integrity of the catheter wall

63. Potential benefits of activated vitamin D therapy include:
A) Decrease of intact parathyroid hormone
B) Decrease of serum calcium
C) Decrease of blood pressure
D) Decrease of arrhythmias

64. For patients receiving an erythropoiesis-stimulating agent (ESA), the new target range for the hemoglobin is:
A) 9g/dL – 11g/dL
B) 12g/dL – 14g/dL
C) No lower limit – 11g/dL
D) 10g/dL – 13g/dL

65. When prescribing an ESA, it is necessary for the transferrin saturation (TSAT) to be at least 20% and the serum ferritin to be:
A) 100 ng/mL
B) 80 ng/mL
C) 50% of normal
D) None of the above

66. Subcutaneous administration of Epoetin alpha is less efficient than IV administration.
A) True
B) False

67. The most common cause of a poor response to ESAs are:
A) Inadequate dosing
B) Poor nutrition
C) Iron deficiency
D) Hyperparathyroidism

68. According to the new regulations, indication(s) for increasing the ESA dose is:
A) If the hemoglobin has not increased by more than 1g/dL after 4 weeks of therapy
B) If the hemoglobin is > 12g/dL
C) If the hemoglobin is > 10g/dL
D) Both A and B

69. According to the new regulations, reducing or interrupting the dose of ESA should be done:
A) If the hemoglobin rises rapidly, e.g., more than 1g/dL in any 2-week period
B) If the hemoglobin level approaches or exceeds 11g/dL
C) If the hemoglobin has not changed
D) Both A & B

70. The desired effect of ESA therapy includes all of the following <u>EXCEPT</u>:
A) An improved quality of life
B) Utilization of iron stores
C) Maintaining the Hct 33-36%
D) Eliminating the need for transfusions

71. An increase in the patient's Hct with the administration of an ESA will cause the clearance of creatinine to:
A) Increase
B) Decrease
C) Normalize
D) There will be no change

72. The mechanism that causes an increase in blood pressure in some patients treated with an ESA may involve:
A) An increase in blood viscosity as the hematocrit decreases
B) An increase in blood viscosity as the hematocrit increases
C) A decrease in blood viscosity as the hematocrit decreases
D) A decrease in blood viscosity as the hematocrit increases

Section XII
NUTRITION

1. A contributing factor to protein malnourishment in renal failure patients is:
   A) Loss of protein in dialysate
   B) Poor dietary protein intake
   C) Catabolic process such as infection
   D) All of the above

2. The recommended protein intake/day for the hemodialysis patient is:
   A) 0.7 gm/kg/day
   B) 1.5 gm/kg/day
   C) 2.0 gm/kg/day
   D) 1.2 gm/kg/day

3. The recommended protein intake/day for the peritoneal dialysis patient is:
   A) 0.7-1.0 gm/kg/day
   B) 1.2-1.5 gm/kg/day
   C) 2.0 gm/kg/day
   D) 1.0 gm/kg/day

4. The best sources of high biological proteins for the renal failure patient are:
   A) Eggs and meat
   B) Milk and beans
   C) Yogurt and peanut butter
   D) Vegetables and eggs

5. The usual recommended sodium restriction for a hemodialysis patient is:
   A) 1 gm/day
   B) 2-3 gm/day
   C) No restriction
   D) 3-4 gm/day

6. With regards to sodium restriction, patients should be taught to:
   A) Not add salt at the table
   B) Use other spices to add flavor to food
   C) Avoid canned and cured foods
   D) All of the above

7. Peritoneal dialysis patients are encouraged to eat larger amounts of protein than hemodialysis patients because:
   A) They usually eat better
   B) They lose protein through the peritoneum
   C) Peritoneal dialysis delivers a higher KT/V than hemodialysis
   D) None of the above

8. Food that is eaten takes 7-9 hours to be digested fully through the tract, this is called:
A) Gastrointestinal lag
B) Peristalsis
C) Slow gut
D) None of the above

9. The recommended protein intake for a child is higher than that of an adult receiving hemodialysis or peritoneal dialysis. For a child less than 30 kg, the recommendation is 1.5-2.0 gm/kg/day. Considering children are smaller in size than adults, why would they require more protein per kg of weight?
I. It is a good source of nutrients because they eat so little of any food
II. Children have a higher metabolic rate
III. Children are more prone to infection; extra protein would help
IV. Proteins are essential to support growth
A) II and IV only
B) I, III, and IV only
C) III and IV only
D) All of the above

10. Inadequate nutrition would be considered at a serum albumin of:
A) < 6.0 gm/dL
B) < 3.5 gm/dL
C) < 5.5 gm/dL
D) < 4.5 gm/dL

11. Which of the following has the lowest potassium content?
A) Potato
B) Banana
C) Peanut butter
D) Pasta

12. The juice most preferred for a diabetic person on hemodialysis is:
A) Orange
B) Apple
C) Pineapple
D) Grape

13. The nutritional supplement formulated especially for the ESRD patient is:
A) Nepro
B) Isocal
C) Ensure
D) Magnacal

14. Intake of whole grain foods are generally <u>NOT</u> recommended for CKD Stage 5 patients receiving dialysis.
   A) True
   B) False

15. The primary indicator of adequate nutritional status is:
   A) Potassium
   B) Albumin
   C) Hematocrit
   D) Iron

16. The recommended fluid intake per 24 hour for the anuric patient on hemodialysis is:
   A) 2 liters of fluid daily
   B) 50 ounces of fluid daily
   C) 32 ounces of fluid daily
   D) All they can drink

17. The following is <u>NOT</u> considered a high source of sodium:
   A) Olives
   B) Soy sauce
   C) Bacon
   D) Garlic

18. Of the following the best snack for a hemodialysis patient is:
   A) Potato chips
   B) Life savers candy
   C) Salted pretzels
   D) Chocolate bar

19. Which of the following medications is <u>NOT</u> regularly prescribed for ESRD patients?
   A) Vitamin E
   B) Iron
   C) Multivitamin
   D) Calcium

20. When instructing a patient about potassium intake, the practitioner should include an explanation regarding:
   A) The use of salt substitutes in place of salt
   B) Eliminating fruits and vegetables: green beans, corn, cherries, apples
   C) Increasing intake of protein foods
   D) Avoiding foods such as chocolate, nuts, and coconut

21. In peritoneal dialysis, dietary caloric requirements:
    A) Increase due to protein loss
    B) Decrease to prevent excessive weight gain and obesity
    C) Decrease with peritonitis
    D) Increase with exchange concentrations

22. The plasma protein commonly used to evaluate the nutritional status of a patient is:
    A) Protein
    B) Albumin
    C) Blood urea nitrogen
    D) Fibrinogen

23. A growing number of healthcare organizations use this measurement to assess nutritional status, rather than a chart of idealized body weight:
    A) Body Mass Index
    B) Dry weight
    C) Lean body weight
    D) None of the above

24. A unit of measure of heat and energy is called:
    A) An amino acid
    B) A protein
    C) A calorie
    D) A carbohydrate

25. The suggested body mass index (BMI) range for normal adults is:
    A) 25-29
    B) 19-24
    C) Less than 19
    D) Over 29

26. In order to decrease the amount of phosphorous in the dialysis patient's diet, the patient should be encouraged to:
    A) Increase his intake of milk and yogurt, especially at breakfast
    B) Substitute fruit punches or "coolers" for colas or drinks at snack time
    C) Use processed "enhanced meat" products in sandwiches
    D) Boil regular meat for lunch and supper meals

27. Constipation is a common problem of dialysis patients because of:
    A) Iron supplementation
    B) Low sodium intake
    C) Slowed gastrointestinal motility
    D) High intake of fibrous foods

28. Patients with renal failure are at risk of malnutrition because:
A) Uremia frequently causes anorexia, nausea and vomiting
B) Uremia causes decreased absorption of water soluble vitamins
C) Many of them experience depression
D) All of the above

29. Suggestions to assist patients in complying with fluid restrictions include:
A) Drinking small amounts of fluid at 1-2 hour intervals
B) Using small cups & glasses for drinking
C) Limiting phosphorous intake
D) Eating more fruits & vegetables

Section XIII
INFECTION CONTROL

1. Which of the following are possible portals of entry for pathogens?

   I Respiratory tract
   II Skin
   III Gastrointestinal tract
   IV Genitourinary tract

   A) I only
   B) I, III, and IV only
   C) II and IV only
   D) All of the above

2. All of the following are modes of transmission of pathogens EXCEPT:
   A) Direct contact
   B) Unsterile dialysate
   C) Surface contact
   D) Aerosolization

3. Isolation of patients with HIV is not necessary because HIV is:
   A) Only transmitted sexually
   B) Not transmitted through the dialysis machine
   C) Not viable outside of the body
   D) B and C

4. Healthcare providers in a dialysis unit can minimize the risk of being infected with blood borne pathogens by:
   A) Proper handwashing techniques
   B) Using universal precautions
   C) Proper disposal of sharps
   D) All of the above

5. Standard Universal Precautions are a :
   A) Practice of infection control that assumes that every direct contact with body fluid is potentially infectious
   B) Practice of infection control that isolates all potentially infectious patients
   C) System which test all patients for HIV and Hepatitis
   D) All of the above

6. Tuberculosis is spread:
   A) By droplets from coughing, sneezing, and speaking
   B) From a draining wound infected with the TB organism
   C) From contact with a body fluid
   D) All of the above

7. Rooms used for TB patients are:
   A) Negative pressure rooms
   B) Positive air flow
   C) Positive air flow, 6 exchanges/min
   D Negative air flow, 6 exchanges/min

8. Ways to handle chemicals safely are:
   A) Wash hands before eating, drinking and smoking
   B) Wear protective clothing
   C) Don't eat, drink, or smoke in work area
   D) All of the above

9. Risk to hazardous chemicals can be reduced by:
   A) Decreasing time exposed
   B) Decreasing the amount of chemicals in the air
   C) Using PPE (Personal Protective Equipment)
   D) All of the above

10. The following items should be discarded as regulated medical waste in red bags:
   A) Blood and blood products
   B) Human pathological waste greater than 20cc
   C) Needles and sharps
   D) A and B only

11. CDC/OSHA's defined exposure to blood/body fluids includes all of the following <u>EXCEPT</u>:
   A) Percutaneous injury (e.g. needle stick, bite)
   B) Contact with urine, feces and tears
   C) Contact with mucous membrane
   D) Contact with skin, when the exposed skin is chapped, abraded, or afflicted with dermatitis

12. Protective eye covering in addition to a mask, gloves and a gown are recommended when the following occurs:
   A) Extensive contact with blood or other body fluid
   B) Aerosolization
   C) Splashing
   D) All of the above

13. After a needle stick exposure the following steps should be initiated:
   A) Notify your supervisor
   B) Wash site with soap and running water
   C) Call employee health/needle stick coordinator
   D) All of the above

14. Infections that most commonly invade the immune system of the renal failure patient are gram-positive organisms. The antibiotic of choice for this type of organism would be:
 A) Gentamycin
 B) Tobramycin
 C) Fortaz
 D) Vancomycin

15. Your patient has an infection at the exit site of his dual lumen catheter that has been reported to be sensitive to Gentamycin. The type of organism likely to be identified as the causative agent is a:
 A) Gram positive organism
 B) Fungus
 C) Gram negative organism
 D) Parasite

16. A caregiver can differentiate between a pyrogenic reaction and an ongoing infection by:
 A) A pyrogenic reaction will result in fever > 101° F
 B) An infection will cause a fever early in treatment
 C) A Pyrogenic reaction will usually occur during the first hour of dialysis
 D) An infection will result in a positive blood culture

17. The most common mode of Hepatitis B transmission is:
 A) Blood leak
 B) Needle stick
 C) Blood transfusions
 D) Surface contact

18. The reason why transmission of Hepatitis B can occur easily in a hemodialysis unit is:
 A) Frequent administration of blood transfusions
 B) Accidental needle stick when removing needles
 C) Use of high flux dialyzers and frequent blood leaks
 D) Hepatitis B virus is viable for 7 days on surfaces

19. A preventative measure(s) that can be practiced in dialysis units to control the incidence of Hepatitis B transmission is (are):
 A) Regular screening of patients and staff
 B) Offering the Hepatitis B vaccine to all patients and staff
 C) Designated area for patients with known HbsAg positivity
 D) All of the above

20. The recommendation to prevent the transmission of HIV includes all of the following practices EXCEPT:
    A) Universal precautions
    B) No recapping of needles
    C) Designated area and equipment for known carriers
    D) Good hand washing

21. The AIDS Confidentiality Law includes all of the following EXCEPT:
    A) Physicians may not order HIV testing without an informed consent
    B) All patients must receive pre and post test counseling when tested for HIV
    C) Information regarding patients AIDS/HIV status must be kept strictly confidential
    D) Information will be shared with everyone

22 The purpose of informed consent before HIV testing includes the following:
    A) To protect the patient against discrimination
    B) To provide information regarding the nature of HIV infection
    C) To discuss safe sexual and lifestyle practices
    D) All of the above

23. If a person has a positive HIV antibody test, all of the following are true EXCEPT:
    A) Medical care and treatment can be provided to slow down the illness
    B) There is no possibility (in pregnancy) of the baby contracting the virus
    C) Resources are available to assist with the emotional aspects of dealing with test results
    D) Contacts should be notified to prevent transmission

24. The major communicable blood-borne pathogens seen in a dialysis unit include all of the following EXCEPT:
    A) Hepatitis B Virus (HBV)
    B) Hepatitis C Virus
    C) Human immunodeficiency virus (HIV)
    D) Tuberculosis

25. Of substantial concern to the dialysis community is the recent development of antibiotic resistant bacteria to all of the following bacteria EXCEPT:
    A) E. coli
    B) Vancomycin resistant enterococci (VRE)
    C) Methicillin resistant staphylococcus aureus (MRSA)
    D) Vancomycin intermediately resistant to staphylococcus aureus (VISA)

26. A method used to prevent bacterial contamination of an area is:
A) Sterile technique
B) Aseptic technique
C) Standard precautions
D) Barrier precaution

27. The major routes of transmission for Viral Hepatitis type B is:
A) Fecal-oral
B) Parenteral only
C) Blood and body fluids
D) All of the above

28. The best way to dispose of used blood tubing, dialyzers, and other blood containing components is to:

A) Double bag them, mark the bag "contaminated" and remove it from the unit as soon as possible
B) Place them in an impervious trash bag marked "biohazardous" and follow institutional policy for disposal of hazardous waste
C) Bag them and place them in a trash can for daily disposal
D) Place them in a trash can and dispose of them daily

29. The major mode of transmission of hepatitis C is:
A) Fecal-oral
B) Blood borne
C) Aerosolization
D) Respiratory secretions

30. The Occupational Safety and Health Administration (OSHA) has recommended that when there is exposure or there is potential for exposure to blood or blood products, the following should be done:
A) Use sterile technique
B) Use universal precautions
C) Use common sense
D) Use gloves

31. The primary source of risk for hemodialysis staff and patients for HIV is:
A) Blood
B) Respiratory secretions
C) Saliva
D) None of the above

32. All of the following statements about hepatitis B vaccine are correct EXCEPT:
A) The vaccine provides passive hepatitis B immunity
B) Hepatitis B vaccines are given in a three or four dose series
C) Serologic testing should be performed 1-2 months after administration of the last dose of the vaccine series
D) Patients with anti-HBs level of less than 10 mIU/mL after the vaccine series should be revaccinated followed by serologic testing

33. When dialysis machinery is being operated, the MOST important reason to promptly wipe up spilled fluid from the machinery or floor is to protect the:
A) Other dialysis equipment from contamination
B) Staff from accidents due to slipping and from shock hazards
C) Environment from growth of organisms
D) Machinery parts from corroding

34. Before you initiate any dialysis procedure, you MUST wash your hands, because handwashing:
A) Prevents skin bacteria from multiplying
B) Kills endotoxins on the skin surface
C) Removes deep-surface bacteria
D) Removes potentially pathogenic skin organisms

35. The dialysis staff should not smoke, eat, or drink in the dialysis treatment area or laboratory primarily because doing so:
A) Causes unpleasant odors
B) Is unprofessional
C) May spread infection
D) Is upsetting to patients

36. Hepatitis B virus in a dialysis facility is most likely to be transmitted by:
A) Contamination of surfaces
B) Transfusion of infected blood
C) Airborne transmission
D) Aerosolization of blood

37. The approach recommended by the CDC and referred to as "universal blood and body fluid precautions" includes all of the following EXCEPT that:
A) Healthcare workers who have dermatitis should put a protective dressing over wounds before performing direct patient care
B) To prevent needle-stick injuries, needles should not be recapped or otherwise manipulated by hand
C) Mouthpieces, resuscitation bags or other ventilation devices should be available for use in areas in which the need for resuscitation is predictable
D) Gloves should be worn for exposure to blood and body fluids, mucous membranes or non-intact skin of all patients

38. A positive reaction to a Mantoux test (PPD) is characterized by:
A) A soft, regular non-indurated area
B) Pruritis at the injection site
C) Erythema at the injection site
D) A firm, irregular area of induration

39. Aseptic technique is required for all the following EXCEPT:
A) Attaching a syringe tip to a needle
B) Attaching the dialysis fluid hoses to the dialyzer
C) Attaching the heparin syringe to the heparin line
D) Attaching the end of the bloodline to the end of the access cannula line

40. Important components of a hand cleansing program include all EXCEPT:
A) Wash hands when entering or leaving patient care areas
B) Wash hands before gloving & after glove removal
C) Wash hands after touching an environmental surface
D) Wash hands with soap & water for five seconds

41. These actions should be taken when an occupational exposure to blood occur:
I Wash needle sticks with soap & water
II Flush splashes to mouth, nose or skin
III Irrigate eyes with clean water, saline or sterile saline solution
IV Report to supervisor
A) I, II, III
B) I, II, IV
C) II, III, IV
D) I, II, III, IV

42. The most common route of transmission of MRSA is from:
A) Contaminated surface areas
B) Contaminated healthcare worker's hand
C) An infected carrier patient
D) A carrier healthcare worker's nose

43. Redness and hard swelling at the site of a mantoux test means that a person:
A) Was exposed to TB
B) Was exposed to hepatitis
C) Has shingles
D) Has HIV

44. One of the facts that contribute to the development of drug-resistant organisms is:
A) Non-adherence to handwashing procedures
B) Not completing the full course of antibiotics as prescribed
C) Prescribing antibiotics based on culture & sensitivity reports
D) Use of antibiotics to treat viral infections

45. Patients with CKD should receive a booster dose of hepatitis B vaccine when their hepatitis B antibody titers (anti-HBs) drop to:
A) < 15 mIu/mL
B) < 5 mIu/mL
C) < 10 mIu/mL
D) < 20 mIu/mL

46. Hand hygiene refers to:
A) Handwashing using plain soap
B) Using an antiseptic hand rub (e.g. alcohol, chlorhexidine, iodine)
C) Handwashing using antimicrobial soap and water
D) All of the above

47. Hand hygiene adherence in health-care facilities might be improved by:
A) Providing personnel with individual containers of alcohol-based hand rubs
B) Providing personnel with hand lotions or creams
C) Providing personnel with feedback regarding hand hygiene adherence/performance
D) All of the above

48. The most common mode of transmission of microorganisms is (are):
A) Healthcare worker's hands to patients
B) Patient to patient
C) Patient to healthcare giver
D) All of the above

49. Gloves should be changed:
A) Whenever dirty
B) When moving from contaminated areas to clean areas
C) When moving between patients
D) All of the above

50. Alcohol-based hand rubs have good or excellent antimicrobial activity against all of the following <u>EXCEPT</u>:
A) Viruses
B) Fungi
C) Mycobacteria
D) Bacterial Spores
E) Gram-positive and gram-negative bacteria

51. Alcohol-based hand rubs are indicated for all of the following clinical situations EXCEPT:
   A) When the hands are visibly soiled
   B) Preoperative cleaning of hands by surgical personnel
   C) Before inserting urinary catheters, intravascular catheters or other invasive devices
   D) After removing gloves

52. The following statements regarding hand hygiene in health-care settings are true EXCEPT:
   A) Overall adherence among health-care personnel is approximately 40%
   B) Poor adherence to hand-hygiene practice is a primary contributor to health-care associated infection and transmission of antimicrobial-resistant pathogens
   C) Personnel wearing artificial nails or extenders have been linked to nosocominal outbreaks
   D) Hand hygiene is not necessary if gloves are worn

53. Hands should be washed:
   A) Before putting on and after removing gloves
   B) Before and after and between patient contacts
   C) After exposure to blood, body fluids, secretions, excretions or contaminated items
   D) Before entering and upon exiting the dialysis treatment area
   E) All of the above

Section XIV
PERITONEAL DIALYSIS

1. Fluid removal is enhanced in peritoneal dialysis (PD) by:
   A) Applying positive pressure and extending diffusion time
   B) Applying negative pressure and shortening diffusion time
   C) Decreasing dextrose concentration
   D) Increasing the dextrose concentration

2. Clearance of urea and creatinine in PD is:
   A) Equal to hemodialysis
   B) Twice as efficient as hemodialysis
   C) 4 to 5 times less efficient than hemodialysis
   D) None of the above

3. The following factors or conditions reduce the efficiency of PD:
   - I. Peripheral vascular disease
   - II. Multiple episodes of peritonitis
   - III. Cool dialysis solution
   - IV. Hypertonic solutions with greater than 2.5 mg/dL of dextrose

   A) I and II only
   B) II, III, and IV only
   C) I, II, and III only
   D) All of the above

4. The national standard for minimal delivered dose of total small solute clearance in peritoneal dialysis to minimize morbidity/mortality rates should be a Kt/V of at least:
   A) 0.8/day
   B) 1.2/week
   C) 1.7/week
   D) 4.0/week

5. All of the following include methods for changing the dialysis prescription of a patient receiving PD <u>EXCEPT</u>:
   A) Dwell time
   B) Drain time
   C) Fill volume
   D) Number of exchanges

6. The method(s) that can be utilized to affect the ultrafiltration of a patient on PD is (are):
   A) Concentration of dextrose in dialysate
   B) Dwell time
   C) Drain time
   D) All of the above

7. The dialysate utilized for PD comes in several concentrations. Among them is a 1.5% dextrose. This equals:
   A) 2.27 g/dL of dextrose
   B) 3.18 g/dL of dextrose
   C) 1.38 g/dL of dextrose
   D) 3.86 g/dL of dextrose

8. The type of organisms that commonly cause peritonitis are:
   A) Fungal
   B) Gram negative bacteria
   C) Gram positive bacteria
   D) Viral

9. Peritonitis is often caused by organisms found:
   A) In the eyes and mouth
   B) In the nose and on the hands
   C) On the exit site and the tunnel
   D) All of the above

10. Peritonitis can effectively be treated with antibiotics either intravenously or intraperitoneally.
    A) True
    B) False

11. Mr. T is receiving PD and develops a fungal infection. The doctor will most likely order Amphotericin to be administered to treat the peritonitis because it is indicated for the treatment of fungal infections.
    A) True
    B) False

12. Effluent drained from the peritoneal cavity should be monitored for:
    - I. Color
    - II. Volume
    - III. Fibrin
    - IV. Clarity

    A) I and II only
    B) II and III only
    C) II and IV only
    D) All of the above

13. The following must be considered when determining a PD prescription:
    A) Size of the patient
    B) Peritoneal equilibration test (PET) results
    C) Lab values
    D) All of the above

14. Infusing cold dialysate into the peritoneal cavity can:
    A) Cause chills and cramps
    B) Cause vasoconstriction
    C) Decrease or inhibit clearance
    D) All of the above

15. Patients receiving PD require dialysis and adequate nutrition to improve morbidity and mortality rates. The dialysis prescription should be adjusted when the:
    I. Weekly KT/V of urea falls below 1.7
    II. nPNA (normalized protein equivalent of total nitrogen appearance) falls below 0.8 gm/kg/day
    III. Renal urea clearance approximates 7 ml/min or less
    A) I and II only
    B) II and III only
    C) I and III only
    D) All of the above

16. Intraperitoneal administration of Epoetin alpha must be placed into a dry peritoneal cavity or one with a minimal amount of dialysate because:
    I. Dialysate will dilute the medication thus slowing its absorption
    II. Administration into a dry cavity improves absorption
    III. Epoetin alpha should not be given intraperitoneally
    A) I and II only
    B) I and III only
    C) II and III only
    D) All of the above

17. Ultrafiltration in PD is accomplished by the utilization of:
    A) Hypertonic dialysate
    B) Hypotonic dialysate
    C) Isotonic dialysate
    D) None of the above

18. Abdominal pain, fever, and a cloudy effluent in CAPD:
    A) Strongly suggest infectious peritonitis
    B) Reflect showers of cholesterol crystals causing peritoneal emboli
    C) Occur periodically due to peritoneal irritation requiring no treatment
    D) Are reason to stop CAPD and change the catheter

19. The peritoneal membrane is:
    A) An impermeable membrane
    B) Continuous in females
    C) One to two square meters and approximates body surface area
    D) A membrane that lines the abdominal and thoracic cavities

20. The most common complication of peritoneal dialysis is:
   A) Peritonitis
   B) Pain on inflow
   C) Protein loss
   D) Hypokalemia

21. In diabetic PD patients:
   A) Glucose control can be managed by adding insulin to the dialysate
   B) Diabetes accelerates due to high glucose concentration in the dialysate
   C) Glycosolated hemoglobin concentrations are higher than in diabetics on hemodialysis
   D) Insulin requirements are similar to diabetic hemodialysis patients

22. Cyclers used for peritoneal dialysis are automatic exchange devices which:
   A) Mix water and electrolytes for hemodialysis
   B) Do plasma exchange procedures
   C) Offer patients exercise while on hemodialysis
   D) Regulate instillation and drainage of the dialysis solution

23. The dextrose load from peritoneal dialysis solution can cause:
   A) Hypoglycemia
   B) Hyperglycemia
   C) Weight loss
   D) Hypocalcemia

24. Approximately 25% of all causes of peritonitis are:
   A) Gram positive bacteria
   B) Gram negative bacteria
   C) Fungi
   D) Virus

25. During peritoneal dialysis, the nurse must continually evaluate the patient for retention of dialysate. This complication would be indicated by:
   A) Fractional urine of 0.5%
   B) Return of fecal material in the outflow
   C) An increase in sodium transfer to serum
   D) Outflows of less than 100mL of inflow

26. Signs and symptoms of exit site infection include:
   A) Epidermis progressing into the sinus tract
   B) Plain granulation tissue in the sinus tract
   C) Any external drainage
   D) Dark, purplish discoloration of the exit site

27. Staphylococci cultured from the nasal carriage has been shown to be a major risk factor in the development of _____ in the PD patient.
A) Pneumonia
B) Exit site infection
C) Tuberculosis
D) Urinary tract infection

28. The following sign(s) and symptom(s) are likely in a patient who has developed peritonitits.
A) Abdominal pain
B) Flu-like symptoms (nausea, vomiting, diarrhea)
C) Fever and chills
D) All of the above

29. The peritoneal membrane:
A) Lines the abdominal wall and abdominal organs
B) Is 2-3 times the body surface area in size
C) Is composed of several layers of mesothelial cells
D) Is a non-permeable membrane

30. The transport of a particular solute across the peritoneal membrane is influenced by:
A) Blood flow to the peritoneal membrane
B) Permeability of the peritoneal membrane
C) Concentration of the solute in the blood and dialysate
D) All of the above

31. Diffusion during peritoneal dialysis can be enhanced by all BUT one of the following:
A) Increasing the dialysis flow
B) Increasing the blood flow to the peritoneum
C) Adding heparin to the dialysis solution
D) Maintaining a high concentration gradient

32. Substances that are routinely lost in the dialysate effluent include:
A) Proteins and amino acids
B) Red blood cells
C) Water soluble vitamins
D) All BUT B

33. Phases of peritoneal dialysis exchange include:
A) Inflow (or infusion)
B) Diffusion (also called dwell or equilibration)
C) Reserve (or repopulation)
D) All EXCEPT C

34. Dialysis solution is warmed to body temperature in order to:
A) Enhance clearance
B) Prevent vasoconstriction
C) Prevent chilling and/or a decrease core body temperature
D) All of the above

35. A key element in the prevention of peritonitis is:
A) Use of aseptic technique
B) A high protein diet
C) Reduced frequency of high dextrose exchange
D) Transfer to CAPD from PD

36. Poor dialysate outflow might be due to all <u>EXCEPT</u> which of the following?
A) Kinked tubing
B) Addition of medications to the dialysis solution
C) Mal-positioned catheter
D) Excessive fibrin

37. Fluid removal or ultrafiltration in peritoneal dialysis is an osmotic loss and may be enhanced by:
A) Lengthening dwell time
B) Increasing dextrose concentration in dialysis solution
C) Limiting low quality protein intake
D) Less frequent exchanges

38. Exit site inflammation can result in all but which one of the following?
A) Tunnel infection
B) Abdominal wall cellulitis
C) Fibrin build-up within the catheter
D) Peritonitis

Section XV
OTHER TREATMENT MODALITIES

1. In the first year following transplant, the MOST frequent cause of death is:
   A) Infection
   B) Cardiovascular accident
   C) Pulmonary embolus
   D) Rejection

2. Continuous arteriovenous hemofiltration or hemodialysis CAVD/D:
   A) Is a low transfer rate solute extraction technique especially suited to unstable patients in an intensive care unit
   B) Offers advantage in intensive care because of the minimal need for nursing supervision
   C) Can be adopted for maintenance therapy in stable, ambulatory ESRD patients
   D) Avoids the need for and risk of anticoagulation

3. Hemofiltration is considered a substitute for hemodialysis in the treatment of ESRD. This mode of treatment involves all of the following EXCEPT:
   A) The ultrafiltrate during the procedure is discarded and the volume is partially replaced by a sterile, Pyrogen-free, physiologic substitution fluid
   B) The rate of hemofiltration is determined by transmembrane pressure, surface area, and transport properties of the membrane, blood viscosity and hematocrit
   C) Solutes are removed by the use of convective transport through a semi-permeable membrane, similar to that of the glomerular membrane of the normal kidney
   D) Solutes are moved by the use of diffusive transport, dependent on the concentration gradient across a semipermeable membrane into a dialysate bath

4. During continuous arterio-venous hemofilter (CAVH) solutes are removed by:
   A) Convection
   B) Diffusion
   C) Osmosis
   D) Hemoperfusion

5. During continuous arterio-venous hemodialysis (CAVHD) solutes are removed by:
   - I. Diffusion
   - II. Osmosis
   - III. Convection
   - IV. Back filtration

   A) 1 and 11 only
   B) 1 and 111 only
   C) 1,11 and 111 only
   D) All of the above

6. The major contributing factor to the complication of infection in a transplant patient is:
   A) Low resistance to infection
   B) Surgical wound has not yet healed
   C) Immunosuppressive medication
   D) Poor hygiene

7. Charcoal hemoperfusion is used for:
   A) Drug overdose
   B) Elevated ammonia
   C) Elevated BUN
   D) Aluminum toxicity

8. A key technical point to keep in mind in dialyzing a patient who has overdosed on medications is that
   A) Percutaneous cannulation of a large central vein using a dialysis catheter will be required
   B) High-flux, high efficiency dialyzers with high urea clearance should be used if possible
   C) A single 3-hour treatment will substantially lower the blood levels of most poisons for which hemoperfusion is effective
   D) All of the above

9. Which one of the following modalities would be <u>MOST</u> effective for an ESRD patient requiring immediate treatment after taking an overdose of potassium?
   A) Peritoneal dialysis
   B) Gastric lavage
   C) Hemoperfusion
   D) Hemodialysis

10. Which of the following statements about short daily hemodialysis (SDHD) is accurate?
    A) It is more difficult to remove urea when SDHD is employed
    B) SDHD is usually performed six times a week for approximately 1.5 to 3 hours per day
    C) Dialysate flow rates are usually lower (≈200mL/min) than those of conventional hemodialysis
    D) The minimum standard Kt/V value will be in the range of 4.0 to 5.0

11. The most frequent problem associated with hypotension during CRRT is:
    A) Imminent cardiac arrest
    B) Clotting of the hemofilter
    C) Increased ultrafiltration
    D) Electrolyte imbalances

12. A patient with ESRD who refuses temporary CRRT secondary to medicinal instability generally will need which therapy?
   A) SCUF
   B) CAVH
   C) CAVHD
   D) CAVU

13. Which of the below are indications for CRRT?
   A) Fluid and solute removal
   B) Hypertension and hyperkalemia
   C) Hyperalimentation and low urine output
   D) Cardiac bypass surgery and sepsis

Section XVI
MISCELLANEOUS

1. The best way to ensure that your documentation on the patient's flow sheet is complete and accurate is to:
   A) Document promptly at the end of your shift
   B) Do your charting as soon as possible
   C) Keep notes at the patient's bedside and chart them at the end of the patient's treatment
   D) Make notes throughout the day on worksheets, then refer to them as soon as you document at the end of your shift

2. The dialysis technician has noticed that a maintenance person who is working around the unit has an open cut on one hand. The technician should:
   A) Review precautions regarding infections with the worker
   B) Report the incident to the maintenance supervisor
   C) Report the incident to the head nurse
   D) Explain why this worker should request assignment elsewhere until the cut is healed

3. The following records are important for a dialysis program to maintain and have available for inspecting agencies:
   A) Water cultures & LAL results
   B) Dialysis machine maintenance records
   C) Employee performance appraisals
   D) All of the above

4. A patient complains to you that during his most recent dialysis he underwent a great deal of pain due to your new co-worker's venipuncture technique. Your most constructive response to this patient would be to:
   A) Inform the patient that this technician is new and is sure to improve with practice
   B) Tell the new co-worker of the patient's complaint and suggest that she request to be removed from this patient's treatment
   C) Listen to the patient's concerns and to inform the head nurse
   D) Inform the patient that he has the right to refuse venipuncture by the new technician

5. The conservative management of a patient with chronic renal failure involves:
   A) Dietary management, medications and frequent medical follow-ups
   B) Dialysis, dietary restrictions and medications
   C) Chronic peritoneal dialysis
   D) Intermittent chronic dialysis, as dictated by symptoms

6. The technician's role is:
   A) Strictly regulated by the Federal Register
   B) Defined by the Nurse Practice Act
   C) Defined by individual states
   D) Defined by the Medical Practice Act

7. The primary criterion for selecting a continuing education program is the:
   A) Lecturer's recognition in the field
   B) Content relevance to work
   C) Recommendation of a peer
   D) Number of co-workers attending the program

8. Which one of the following is <u>NOT</u> essential to the successful completion of a research project?
   A) Review of the literature in the problem area
   B) Publication of results
   C) Problem definition
   D) Obtaining permission of the unit manager and the study subjects

9. The professional organization for dialysis technicians is:
   A) The Board of Nephrology Nurses and Technologists (BONENT)
   B) The American Association for Nephrology Nurses and Technicians (AANNT)
   C) The National Association of Nephrology Technologists (NANT)
   D) The National Organization for Patient Care Technicians (NOPCT)

10. Detailed charting:
    A) Wastes effort
    B) Wastes money
    C) Costs you valuable time you need to give quality patient care
    D) Ensures continuity of care

11. Which of the following statements is correct about abbreviating?
    A) Every facility should have an approved abbreviation list
    B) Creating abbreviations saves time for the reader
    C) Abbreviating drug names and dosages helps reduce medication errors
    D) Writing out questionable abbreviations could make a jury think you are hiding something

12. On which part of the chest is the heel of the hand placed in order to perform chest compressions in cardiopulmonary resuscitation (CPR)?
    A) The upper third of the sternum
    B) The middle of the sternum
    C) The lower third of the sternum
    D) The xiphoid process

13. In performing emergency cardiopulmonary resuscitation, attention should be given to the victim's immediate needs in the following order:
    A) Airway, breathing, and circulation
    B) Circulation, breathing, and airway
    C) Breathing, airway, and circulation
    D) Precordial thump, breathing, and circulation

14. How should information be presented to patients with chronic kidney disease?
    A) In a neutral manner, to let the patient decide what to expect
    B) In a positive manner, emphasizing what can be done and what the patient can do for him or herself
    C) It doesn't really matter, the patient's productive lives are over
    D) In a realistic way, emphasizing all the probable pitfalls

15. Which of the following is NOT a characteristic of adult learners?
    A) Adult learners integrate new information into an existing framework
    B) Adult learners seek out relevant information
    C) Adult learners prefer practical information
    D) Adult learners prefer to learn about theory

16. The principles of learning includes the following:
    I. Consolidating information enhances learning
    II. Repetition positively influences learning
    III. Active learner involvement in the learning process enhances learning
    IV. Using examples facilitates learning
    V. Correcting wrong information immediately facilitates learning
    A) I, II, and III only
    B) II, IV, and V only
    C) I, II, III, and IV only
    D) All of the above

17. Strategies for enhancing learning include:
    A) Organization
    B) Repetition
    C) Specificity
    D) All of the above

18. The legislation which was passed to provide equitable care for all ESRD patients is the:
    A) 1987 Reuse standards
    B) 1972 Social Security Amendment
    C) 1989 Transplant Program
    D) Bill of Rights

19. Agencies that provide services for ESRD patients are:
    - I. American Kidney Fund
    - II. National Kidney Foundation
    - III. American Association of Kidney Patients
    - IV. ESRD Networks

    A) I, II, and III only
    B) II, III, and IV only
    C) II and III only
    D) All of the above

20. An advanced directive includes:
    A) DNR, Health Care Proxy, Living will
    B) Health care proxy, bill of rights, living will
    C) Living will, DNR, bill of rights
    D) DNR, consent form, healthcare proxy

21. The primary responsibility of all healthcare workers is to:
    A) The immediate supervisor
    B) The State Department of Health
    C) The patient
    D) The Medicare program

22. Measures that can be taken in order to protect our back from injury is (are):
    A) Maintain good posture
    B) Sit in chairs with adequate back support
    C) Get close to the object being lifted
    D) All of the above

23. Talking to a colleague in the public cafeteria of a hospital about a specific patient may result in charges of all <u>EXCEPT</u> which one of the following?
    A) Professional misconduct
    B) Unprofessional conduct
    C) Gross incompetence
    D) Violation of confidentiality

24. In the event of a natural disaster (e.g. hurricane, fire, flood, etc.), the primary responsibility of the staff in the dialysis setting is the protection of the:
    A) Dialysis staff
    B) Patients
    C) Dialysis equipment
    D) Facility

25. In a safety-oriented facility, which policy should a safe-patient handling program include?
A) Staff members must adhere to the program as long as staffing is adequate
B) Staff members have the option to participate in the program
C) Caregivers are required to keep a written log of their injuries
D) Caregivers can report injuries without fear of being blamed

26. The focus of continuous quality improvement (CQI) is documentation & resolution of problems.
A) True
B) False

27. The four main purposes of documentation are:
A) Communication, CYOA, performance appraisal, reimbursement
B) Communication, CYOA, performance appraisal, auditing
C) Communication, CYOA, permanent record of events, reimbursement
D) Communication, CYOA, permanent record of events, auditing

28. Documenting that manufacturer guidelines were followed when using dialysis equipment is important primarily to:
A) Check the reliability of the manufacturer's guidelines
B) Complete chart audits
C) Obtain reimbursement
D) Show that all risk was eliminated

29. The discharge note for your dialysis patient should:
A) State that "the patient was stable when discharged"
B) State that "the patient was discharged without incident"
C) Describe the patient's medical prescription at discharge
D) Describe the patient's condition at discharge

30. Your hemodialysis patient states, "I eat anything I want". Your primary reason for documenting this information is to:
A) Communicate to the dietitian a problem with noncompliance
B) Communicate to the staff that this is an uncooperative patient
C) Cover yourself in the event your manager audits the record
D) Cover yourself in the event of a lawsuit

31. Proper documentation by the nurse/technician of the status of the dialysis equipment on the patient record should include:
A) Recording the pre-dialysis formalin results
B) Recording the number of times in the past year the equipment has been serviced
C) Noting the status of safety alarms during the dialysis treatment
D) Noting the time when the equipment is next scheduled for maintenance

Match items in Column A with the correct phrase from Column B

32. _____ < 200 CFU/mL
33. _____ Pyrogenic reactions
34. _____ Bacterial contamination
35. _____ ABC
36. _____ < 2 EU/mL
37. _____ Endotoxins
38. _____ CPR
39. _____ Hemolysis
40. _____ RACE

A) Type of extinguisher for all categories of fire
B) Should be initiated when someone is found pulseless & breathless
C) Acronym for an important safety rule
D) One type of water emergency
E) Caused by endotoxins found in dialysis fluid
F) Caused by dialysis solutions contaminated with chloramines, copper or nitrates
G) Microbial count allowed for water used to prepare dialysate
H) Bacteria contain these substances which can pose a significant risk to dialysis patients
I) Endotoxin level allowed in water or dialysate

41. According to the American Heart Association, when performing CPR in adults, the correct compression-ventilation ratio is ___ compressions and ___ breaths.
A) 15, 2
B) 15, 1
C) 30, 2
D) 20, 1

42. During CPR, rescuers should try to minimize interruptions to less than ___ seconds.
A) 1
B) 10
C) 5
D) 20

43. The correct rate for giving compressions in adult CPR is ___ compressions a minute.
A) 100
B) 80
C) 60
D) 70

44. When administering CPR in the workplace, OSHA requires the healthcare workers use standard precautions. Standard precautions would include the use of:
A) Gloves only
B) Barrier devices or bag-mask systems, gloves and goggles
C) Barrier devices only
D) Mask, gloves and goggles only

45. According to recent statistics, a healthy human brain may survive without oxygen for up to ___ minutes without suffering any permanent damage.
A) 4
B) 5
C) 10
D) 12

46. Before you start CPR, you must remember to check the victim for:
A) Dentures
B) Bleeding
C) Responsiveness
D) Temperature

47. The depth of compressions in adult CPR should be approximately:
A) 1/2 to 1 inch
B) 1 to 1 1/2 inch
C) 1 1/2 to 2 inches
D) 2 to 3 inches

48. You should check the victim for responsiveness by:
A) Using smelling salts
B) Shaking him and shouting, "Are you okay?"
C) Pouring cold water on his face
D) All of the above

49. If the adult victim remains unresponsive, you should:
A) Dial 9-1-1 before starting CPR
B) Start CPR before dialing 9-1-1
C) Wait to see if the victim regains consciousness
D) None of the above

50. To check if the victim is breathing, you should:
A) Listen for exhaled air
B) Watch for his chest to rise and fall
C) Feel for exhaled air
D) All of the above

51. The most common airway obstruction is:
A) Dentures
B) Food
C) The tongue
D) None of the above

52. The technique used to clear the victim's airway is:
A) Lift chin up, tilt head back
B) Push chin down, tilt head forward
C) Lift chin up, turn head sideways
D) None of the above

53. Check for circulation by feeling for a pulse at the:
A) Jugular vein
B) Heart
C) Carotid artery
D) None of the above

54. When assisting the victim with breathing:
A) Pinch the victim's nose closed
B) Do not over-inflate the victim's lungs
C) Allow the victim to exhale on his own
D) All of the above

55. When administering chest compressions:
A) Position your hands on the sternum
B) Give 2 breaths after 30 compressions
C) Apply the "2 hands, 2 inches" rule
D) All of the above

Section XVII
CASE STUDIES

Case Study #1 (Questions 1-3)

Mr. D, a 50 year old male, has an A-V fistula created in preparation for maintenance hemodialysis treatments.

1. The following is reviewed with Mr. D to assist in development of his fistula:
   A) Perform arm exercises for 30-60 minutes, 5-6 times a day
   B) Protect the arm by using a sling
   C) Perform arm exercises by squeezing a rubber ball while compressing the upper arm with the opposite hand
   D) Wrap the arm with an elastic bandage to protect the fistula

2. The physiologic principle behind fistula development is:
   A) Antegrade flow is deprived by retrograde flow
   B) Collateral circulation is diverted toward the newly formed fistula
   C) Arterial blood flow against venous resistance will cause vessels to dilate
   D) All of the above

3. Thrombosis is a common complication of AV accesses. One way to help prevent this from occurring would be:
   A) Maintaining immobility of the arm during dialysis
   B) Avoiding extrinsic pressure in the vessels
   C) Avoiding hypertension during dialysis
   D) Maintaining the patient below his dry weight

Case Study #2 (Questions 4-6)

Ms. M is a 27 year old with SLE who started peritoneal dialysis three months ago. She is very concerned with her weight and does not want to get "fat". However, you notice she has an Albumin of 3.0 and she doesn't understand "what is the big deal about eating protein."

4. What would be the best answer for Ms. M?
   A) Protein helps us stay healthy
   B) Proteins are not really so important because she's already grown to potential
   C) Protein is essential for someone in child bearing age
   D) Proteins are necessary for body tissues, especially our immune system

5. Ms. M also is questioning why she is having a problem now because when she was on hemodialysis she didn't have a problem with her protein intake. You explain:
   A) "You are doing better now and need more protein."
   B) "During peritoneal dialysis, more protein is lost in the dialysate."
   C) "Maybe you weren't eating enough protein and nobody noticed."
   D) None of the above

6. She also mentions that she is worried about getting "fat" because she's noticed her weight has increased and she feels very thirsty. What else would you check in her labs and instruct her to do?
   A) Glucose; avoid concentrated sweets
   B) Triglycerides; avoid fried foods
   C) Sodium; avoid canned foods and adding salt
   D) All of the above

Case Study # 3 (Questions 7-10)

Ms. J has hyperkalemia almost every month when labs are drawn pre-dialysis. When you review the foods she is eating, she admits to eating plantains daily. She "just has to have them," she says. "I can deal with the other restrictions, but this is something I'm having difficulty with."

7. What could she do during food preparation to eliminate some of the potassium from the plantains or from other foods that are high in potassium?
   I. Soak them overnight or boil them and then replace in water
   II. Nothing can be done. She just can't eat plantains.
   III. If she must have them, she should limit them to once a week.
   IV. She could only have them boiled.

   A) I, III, and IV only
   B) III and IV only
   C) I and III only
   D) IV only

8. What other foods should you also discuss with Ms. J, which she might be eating that are also high in potassium and can be prepared similarly?
   A) Potatoes
   B) Rice
   C) Green leafy vegetables
   D) A and C only

9. When the patient is preparing meals, lemon and other spices are recommended to make food more palatable. However, your patient has asked if it would be okay to use a salt substitute such as Lo-salt. What would be your answer?
   A) Yes, salt substitutes also help to make food tastier.
   B) No, salt substitutes are high in potassium.
   C) No, salt substitutes have hidden sodium.
   D) Yes, salt substitutes still have some sodium but less than table salt.

10. Your patient states she often gets heartburn for which she takes Mylanta. But then it causes her to become constipated so then she needs to take Milk of Magnesia (MOM). Is there a problem with a renal failure patient taking either of these medications?
   A) Mylanta is okay as an antacid; it doesn't contain aluminum.
   B) Both medications contain magnesium and can cause hypermagnesemia.
   C) MOM is okay; Mylanta has aluminum.
   D) Both medications are okay, as long as they don't have aluminum.

Case Study # 4 (Questions # 11 and 12)

Ms.S is a 56 year old patient whose ESRD is caused by hypertension. She has been on hemodialysis for 2 weeks and has just completed her treatment of 3 hours on a high flux dialyzer. The target weight loss was 2.5 kgs to reach her dry weight of 53 kgs. During her post-dialysis assessment she stands to get weighed, becomes very dizzy, diaphoretic, and her BP drops to 70/44, 30 mmHg lower than the sitting pressure.

11. What should you do first?
   A) Call the nurse or doctor.
   B) Sit her back down and place in trendelenburg.
   C) Support her so she doesn't fall.
   D) Give broth.

12. There are many possible-contributing factors to this hypotensive episode. A review of Ms. J's treatment records shows that the usual weight removal during dialysis is 3-3.5 kgs; BP standing post dialysis range is 98-102/60-72, and there have been no changes in the treatment and prescription. Another question that needs to be considered is
   A) The size of the dialyzer
   B) An increase in her dry weight
   C) Medications
   D) Food consumption pre-dialysis

Case Study # 5 (Questions 13-14)

Mr. P is a 34-year-old male diagnosed with AIDS, performing home peritoneal dialysis. His primary disease is hypertension and he is currently doing well. He decides to withdraw from dialysis therapy.

13. This situation is difficult for all involved. It is considered to be
    A) A risk to the patient.
    B) A patient right.
    C) An ethical issue.
    D) All of the above.

14. What are some of the factors to consider with Mr. P's decision?
    A) Quality of life for the patient
    B) Competency of the patient
    C) Financial burden to institution
    D) A and B

Case Study # 6 (Questions 15-17)

A new patient from the medical floor arrives for hemodialysis. The patient's BUN is 200, K-5.5, and Ca-8.5. Dialysis is initiated for 3 hours, counter current dialysate flow, 300-blood flow. One hour into treatment, the patient complains of a headache, nausea, vomiting, and begins to have a seizure.

15. What should you suspect?
    A) Dialysis encephalopathy
    B) Chemical reaction
    C) Disequilibrium syndrome
    D) Hypovolemia

16. The appropriate intervention in this situation would be
    A) Saline 200-300cc, Trendelenburg, minimum UFR
    B) Stop dialysis, call for help, protect the patient
    C) Hypertonic, check the drain for residual chemical
    D) Check water for aluminum

17. To avoid such a situation from occurring, the following could be done:
    A) Purify water to remove aluminum
    B) Proper rinsing after chemical disinfection
    C) Co-current, short dialysis, low blood flow
    D) Accurate estimate of dry weight

Use the following formula to answer Questions 18-23.

Formula:

UF goal divided by the # hours per treatment = UFR

UFR divided by TMP = K_{UF}

UF Goal	Total volume to be removed
TMP	Reading taken from the machine
UFR	Ultrafiltration Rate per hour

18. Patient Data:

Hemodialysis time	3 hours	Dialyzer K_{UF}	7
Dry weight	52 kgs	Today's weight	56kgs
Prime & rinseback	300cc		

Patient will receive 100cc of IV fluids/medications during dialysis treatment

Calculate the patient's UFR and TMP

A) UFR = 1466, TMP = 209
B) UFR = 209, TMP = 1466
C) UFR = 1142, TMP = 163
D) UFR = 63, TMP = 942

19. Patient 1 dialyzes on a reprocessed dialyzer. She achieves a UF goal of 4950 and dialyzes for 2 hours with a TMP of 35. What is the K_{UF} of the dialyzer?

A) 20
B) 80
C) 70
D) 50

20. Patient 2 dialyzes on a reprocessed dialyzer. He achieves a UF goal of 10,350, dialyzes for 3.5 hours, and TMP is 47. What is the K_{UF} ?

A) 62
B) 70
C) 90
D) 100

21. Patient 3 dialyzes on a reprocessed dialyzer. This is the eighth reuse of the dialyzer. The patient achieves a UF goal of 1,350, dialyzes for 2.5 hours, and TMP is 5. What is the K_{UF} ?

A) 60
B) 70
C) 90
D) 108

22. The invitro K_{UF} of a dialyzer is 80. Is there a problem with any of the 3 previous situations (Questions 19-21)?

    I. No, some variation is expected.
    II. Yes, the K_{UF} of Patients 1 and 2 are too low.
    III. Yes the K_{UF} of Patient 2 is too low.
    IV. Yes, the K_{UF} of Patient 3 is excessively higher than it should be.

    A) I only
    B) II and III only
    C) III and IV only
    D) II and IV only

Case Study # 7 (Questions 23-28)

Mr X is new to dialysis and is having some difficulty with his diet and fluid restrictions. He has also experienced some problems during dialysis.

23. Mr. X is gaining more than 3 kg of fluid between treatments. To assist him in controlling his weight gain, the following suggestion is made:
    A) Eat ice cream instead of drinking soda between meals
    B) Eat more fruits and vegetables and less meat and pasta
    C) Suck on hard candies or measured ice ships to relieve a dry mouth
    D) Take no liquids after 8pm on the day before dialysis

24. When instructing Mr. X about his potassium intake, the following suggestion is made:
    A) Eliminate fruits and vegetables
    B) Increase intake of protein foods
    C) Include salt substitutes in place of salt
    D) Soak fruits and vegetables in water before cooking

25. During the last hour of dialysis, Mr. X experiences a drop in blood pressure. Actions to prevent hypotension include:
    A) Discontinuing dialysis
    B) Reducing ultrafiltration
    C) Position the patient in semi-Fowler's
    D) Lower the dialysate sodium concentration

26. The conductivity monitor on Mr. X's machine alarms. All of the following conditions could result in hemolysis <u>EXCEPT</u>:
    A) Low negative pressure
    B) High negative pressure
    C) Hypertonic dialysate
    D) Hypotonic dialysate

27. The air detector on Mr. X's machine alarms and foam is visible in the venous drip chamber. The immediate action of the practitioner should be to:
A) Mute the alarm
B) Clamp the venous line
C) Assess the arterial needle placement
D) Tap the venous chamber

28. Mr. X complains of shortness of breath. While calling the nurse, the technician should also place the patient in:
A) Trendelenburg position on his right side
B) Trendelenburg position on his left side
C) A supine position with his legs elevated
D) A sitting position

ANSWER KEY

Section I
Renal A & P

1. A
2. D
3. B
4. A
5. C
6. D
7. B
8. C
9. D
10. D
11. A
12. B
13. A
14. D
15. A
16. D
17. A
18. C
19. A
20. A
21. D
22. B
23. D
24. B
25. D
26. B
27. G
28. H
29. E
30. C
31. D
32. A
33. B
34. C
35. D
36. B
37. B
38. A
39. D
40. B
41. A
42. C
43. D
44. A
45. A
46. A
47. C
48. A
49. A
50. A
51. D
52. D
53. C
54. B
55. C
56. C
57. B
58. C
59. D

Section II
Causes of Renal Failure

1. B
2. B
3. A
4. C
5. B
6. A
7. D
8. D
9. C
10. D
11. B
12. D
13. D
14. B
15. D
16. A
17. D

Section III
Clinical Manifestations

1. A
2. B
3. A
4. A
5. D
6. D
7. D
8. A
9. A
10. D
11. C
12. C
13. B
14. C
15. D
16. B
17. D
18. D
19. D
20. D
21. D
22. D
23. D
24. C
25. D
26. D
27. A
28. D
29. C
30. A
31. C
32. D
33. C
34. B
35. C
36. D
37. A
38. D
39. C
40. B
41. A
42. A
43. B
44. A
45. A
46. A
47. D
48. A
49. B
50. D
51. D

Section IV
Psychosocial

1. D
2. C
3. C
4. D
5. B
6. C
7. D
8. A
9. D
10. A
11. A
12. D
13. B
14. A
15. D
16. D
17. B
18. C
19. D

Section V
Hemodialysis

1. B
2. B
3. D
4. A
5. A
6. D
7. A
8. A
9. D
10. C
11. D
12. C
13. C
14. D
15. D

16. A
17. A
18. A
19. D
20. B
21. B
22. A
23. A
24. D
25. A
26. A
27. A
28. A
29. D
30. A
31. D
32. A
33. B
34. A
35. A
36. D
37. A
38. A
39. B
40. C
41. D
42. B
43. D
44. B
45. C
46. D
47. B
48. D
49. B
50. C
51. D
52. D
53. A
54. A
55. D
56. B
57. D
58. A
59. D
60. D
61. C
62. A
63. D
64. C
65. C
66. A
67. D
68. B
69. C
70. D
71. D
72. D
73. C
74. D
75. C
76. A
77. A
78. D
79. C
80. D
81. B
82. A
83. D
84. B
85. D
86. C
87. D
88. D
89. B
90. A
91. A
92. C
93. A
94. B
95. C
96. D
97. A
98. A
99. B
100. C
101. C
102. B
103. A
104. C
105. D
106. B
107. D
108. D
109. A
110. D
111. D
112. D
113. B
114. A
115. A
116. A
117. A
118. A
119. D
120. D
121. A
122. B
123. B
124. B
125. C
126. C
127. C
128. D
129. D
130. C
131. B
132. D
133. B
134. B
135. B
136. B
137. B
138. A
139. D
140. C
141. C
142. C
143. D
144. C
145. A
146. C
147. A
148. A
149. B
150. C
151. A
152. B
153. A
154. B
155. C
156. B
157. D
158. D
159. C
160. D
161. B
162. A
163. D
164. C
165. A
166. B
167. B
168. D
169. D
170. B
171. A
172. C
173. D
174. A
175. D
176. A
177. C
178. D
179. D
180. D
181. C
182. A
183. A
184. B
185. B
186. D
187. C
188. D
189. C
190. C
191. C
192. A
193. A
194. B
195. A
196. B
197. C
198. D
199. D

200. D
201. D
202. B
203. C
204. C

Section VI
Vascular Access

1. A
2. C
3. D
4. B
5. D
6. D
7. A
8. D
9. C
10. B
11. D
12. D
13. A
14. D
15. D
16. D
17. D
18. C
19. C
20. C
21. A
22. D
23. B
24. A
25. A
26. A
27. A
28. A
29. B
30. B
31. A
32. A
33. B
34. B
35. D
36. A
37. D
38. C
39. B
40. A
41. A
42. C
43. D
44. D
45. B
46. D
47. A
48. A
49. B
50. D
51. C
52. A
53. B
54. C
55. B
56. A
57. A
58. D
59. A
60. A
61. A
62. D
63. B
64. C
65. B
66. D
67. C
68. A
69. C
70. C
71. B
72. D
73. C
74. A
75. A
76. D
77. A
78. C
79. B
80. D
81. D
82. C
83. C
84. A
85. A
86. A
87. A
88. C
89. F

Section VII
Dialyzer Reprocessing

1. C
2. B
3. D
4. B
5. D
6. A
7. B
8. C
9. B
10. A
11. A
12. A
13. A
14. A
15. D
16. D
17. A
18. B
19. C
20. B
21. A
22. A
23. B
24. A
25. C
26. E
27. D
28. A
29. B
30. B
31. C
32. B
33. D
34. B
35. A
36. A
37. D
38. A
39. B
40. D
41. B
42. D
43. D
44. D
45. D
46. B
47. B
48. D
49. B
50. D
51. A
52. C
53. D
54. A
55. D
56. A
57. C
58. A
59. A
60. A
61. D

Section VIII
Water Treatment

1. D
2. A
3. A
4. D
5. A
6. B
7. A
8. A
9. A
10. C
11. C
12. D
13. C
14. A
15. B
16. A
17. B
18. C

19. A
20. B
21. D
22. C
23. D
24. B
25. E
26. F
27. A
28. C
29. B
30. A
31. C
32. B
33. C
34. D
35. A
36. A
37. A
38. C
39. D
40. C
41. C
42. D
43. A
44. B
45. A
46. A
47. A
48. C
49. B
50. B
51. A
52. B
53. A
54. A
55. A
56. D
57. D
58. C
59. A
60. A
61. C
62. A
63. A
64. B
65. D
66. B
67. A
68. A
69. D
70. D
71. B
72. A
73. C
74. D
75. B
76. B
77. C
78. D
79. A
80. F
81. E
82. G
83. D
84. D
85. C
86. A
87. A
88. C
89. D
90. A
91. C
92. B
93. A

Section IX Biomedical/ Chemistry

1. C
2. A
3. A
4. D
5. B
6. A
7. D
8. C
9. E
10. L
11. H
12. G
13. J
14. I
15. F
16. K
17. B
18. A
19. D
20. C
21. F
22. E
23. H
24. G
25. B
26. A
27. I
28. H
29. C
30. F
31. E
32. G
33. J
34. D
35. B
36. D
37. C
38. A
39. B
40. F
41. G
42. E
43. A
44. A
45. D
46. C
47. F
48. C
49. H
50. B
51. A
52. J
53. I
54. K
55. D
56. G
57. E
58. B
59. A
60. A
61. A
62. A
63. A
64. B
65. B
66. C
67. A
68. D
69. B
70. B
71. C
72. C
73. D
74. B
75. D
76. A
77. C
78. C
79. B
80. C
81. A
82. C
83. B
84. C
85. D
86. A
87. A
88. D
89. B
90. C
91. A
92. D
93. A
94. C
95. D
96. A
97. B
98. A
99. B
100. B
101. B
102. A
103. A
104. D
105. B

106. A
107. D
108. B
109. B
110. A
111. D
112. A

Section X
Labs

1. A
2. B
3. A
4. B
5. A
6. B
7. C
8. A
9. B
10. A
11. B
12. A
13. A
14. D
15. A
16. C
17. D
18. C
19. C
20. B
21. D
22. A
23. A
24. D
25. D
26. B
27. C

Section XI
Medications/
Anticoagulation

1. C
2. A
3. C
4. B
5. C
6. B
7. A
8. A
9. B
10. D
11. C
12. B
13. C
14. D
15. B
16. B
17. C
18. B
19. D
20. A
21. C
22. A
23. C
24. D
25. B
26. D
27. D
28. C
29. D
30. C
31. B
32. B
33. C
34. D
35. A
36. B
37. A
38. C
39. A
40. B
41. B
42. C
43. B
44. A
45. B
46. D
47. A
48. C
49. C
50. C
51. D
52. B
53. A
54. A
55. A
56. A
57. C
58. B
59. B
60. A
61. C
62. C
63. A
64. C
65. A
66. B
67. C
68. A
69. D
70. B
71. B
72. B

Section XII
Nutrition

1. D
2. D
3. B
4. A
5. B
6. D
7. B
8. A
9. A
10. B
11. D
12. B
13. A
14. A
15. B
16. C
17. D
18. B
19. A
20. D
21. A
22. B
23. A
24. C
25. B
26. B
27. A
28. D
29. B

Section XIII
Infection
Control

1. D
2. B
3. D
4. D
5. A
6. A
7. D
8. D
9. D
10. D
11. B
12. D
13. D
14. D
15. C
16. C
17. D
18. D
19. D
20. C
21. D
22. D
23. B
24. D
25. A
26. B
27. C
28. B
29. B
30. B
31. A
32. A
33. B
34. D
35. C

36. A
37. A
38. D
39. B
40. D
41. D
42. B
43. A
44. B
45. C
46. D
47. D
48. D
49. D
50. D
51. A
52. D
53. E

Section XIV Peritoneal Dialysis

1. D
2. C
3. C
4. C
5. B
6. D
7. C
8. C
9. B
10. A
11. A
12. D
13. D
14. D
15. D
16. A
17. A
18. A
19. C
20. A
21. A
22. D
23. B
24. B
25. D
26. C
27. B
28. D
29. A
30. D
31. C
32. D
33. D
34. D
35. A
36. B
37. B
38. C

Section XV Other Modalities

1. A
2. A
3. D
4. A
5. B
6. C
7. A
8. D
9. D
10. B
11. B
12. C
13. A

Section XVI Miscellaneous

1. B
2. A
3. D
4. C
5. A
6. C
7. B
8. B
9. C
10. D
11. A
12. C
13. A
14. B
15. D
16. D
17. D
18. B
19. D
20. A
21. C
22. D
23. C
24. B
25. D
26. B
27. C
28. D
29. D
30. A
31. C
32. G
33. E
34. D
35. A
36. I
37. H
38. B
39. F
40. C
41. C
42. B
43. A
44. B
45. A
46. C
47. C
48. B
49. A
50. D
51. C
52. A
53. C
54. D
55. D

Section XVII Case Studies

1. C
2. C
3. B
4. D
5. B
6. D
7. C
8. D
9. B
10. B
11. B
12. C
13. D
14. D
15. C
16. B
17. C
18. A
19. C
20. A
21. D
22. C
23. C
24. D
25. B
26. C
27. B
28. B

REFERENCES AND ADDITIONAL READINGS

"25 Facts about Organ Donation and Transplantation." Translantation. 2008. National Kidney Foundation. 09 Nov 2008 www.kidney.org/transplantation.html

"ANSI/AAMI RD52:2004/A1:2007 & A2:2007 Dialysate for Hemodialysis." Association for the Advancement of Medical Instrumentation. 2008. AAMI. 09 Nov 2008 www.aami.org/publications/standards/dialysis.html

"ANSI/AAMI RD62:2006, Water Treatment Equipment for Hemodialysis Applications." Association for the Advancement of Medical Instrumentation. 2006. AAMI. 08 Nov 2008 www.aami.org/publications/standards/dialysis.html

"Bloodborne Pathogens and needlestick Prevention." Workplace Safety. 2008. Occupational Safety and health Administration. 16 Nov 2008 www.osa.gov/SLTC/bloodbornepathogens/index.html

"Candidate Examination Handbook." 2008. Board of Nephrology Examiners Nursing and Technology. 06 Nov 2008 www.bonent.org/exam_policies/candidate%20handbook.pdf

"Concise 2008 Annual Data Report Atlas of Chronic Kidney Disease and End Stage Renal Disease in the United States." 2008. United States Renal Data System (USRDS). 20 Nov 2008 www.usrds.org/2008/usrds_booklet_08.pdf

Core Curriculum for the Dialysis Technician. 4th. Madison: Medical Education Institute, Inc., 2008.

Counts, C (ed.) Core Curriculum for Nephrology Nursing. 5th. Pitman: ANNA, NJ, 2008.

CPR Standards. Dallas: American Heart Association, 1997-1998.

Curtis,.J. and Varughese,P.,A Manual for Dialysis Technicians, Third Edition 2003, NANT.

Daugirdas, J.T., Blake, P.G. , and Ing, T.S. (eds). Handbook of Dialysis. Philadelphia: Lippincott, Williams, & Wilkins, 2007.

"Dialyzer Reprocessing." Renal Systems Group. 2007. Minntech Corporation. 12 Nov 2008 www.minntech.com/index.html

Hemodialysis Dialyzers. 2008. Baxter International, Inc. 16 Nov 2008 www.baxter.com/products/renal/hemodialysis/sub/dialyzers.html

Kallenbach, J.Z., Gutch, C.E., Stoner, M.H., Corea, A.L. Review of Hemodialysis for Nurses and Dialysis Personnel. 7th ed. St Louis, Elsevier Mosby, 2005.

"Kidney Pictures." Encyclopedia Britannica. 2008. Britannia.com 18 Nov 2008 www.britannica.com/EBchecked/topicart/317358/99764/Diagram-showing-the-location-of-the-kidneys-in-the-abdominal

Molzahn, A.M. and Butera, E. Contemporary Nephrology Nursing Principles and Practice. 2nd. Pitman: Anthony Janetti, Inc., 2006

Nugent, P. and Vitale, B. Fundamentals of Success: Course Review Applying Critical Thinking in Test Taking. Philadephia: F.A. Davis Company, 2004.

"Personal Protective Equipment." Workplace safety. 2008.Occupational safety and Health Administration. 11 Nov 2008 www.osha.gov/SLTC/personalprotectiveequipment/index.html

"Polyflux Dialyzer Family." In-center Hemodialysis Products. 2007. Gambro Lundia AB. 14 Nov 2008. www.usagambro.com/Pages/InfoPage.aspx?id=4254

"RenalPure® Liquid Acid Concentrate for Hemodialysis." 2008. Rockwell Medical Technologies, Inc. 20 Nov 2008 www.rockwellmed.com/acidsoln.html

Rocco, M.V., and Burkart, J.M. "Kidney Outcome Prediction and Evaluation (KOPE) Study: A Prospective Cohort Investigation of Patients Undergoing Hemodialysis Study and Design and Baseline Characteristics." Annals of Epidemiology Vol. 8, Issue 3(1998): 192-200.

"Updated U.S. Public Health Service Guidelines for the Management of Occupational Exposures to HBV, HCB and HIV and Recommendations for Post Exposure Prophylaxis." MMWR. 2001. Centers for Disease control. 02 Nov 2008 www.cdc.gov/mmwr/preview/mmwrhtml/rr5011a1.html

Varughese, P., Arslanian, J. and Andrysiak, P. A Study Guide for Dialysis Technicians. 3rd. NANT, 2006

Test-Taking Tips

Prepare!

- Arrive early for tests
    - Bring all the materials you will need such as pencils, eraser, and a watch. This will help you focus on the task at hand.
- Be comfortable, but alert
    - Choose a good seat and make sure you have enough room to work, maintain comfortable posture, and don't "slouch."
- Stay relaxed and confident
    - Remind yourself that you are well prepared and are going to do well. If you find yourself anxious, take several, slow, deep breaths to relax.
    - Don't study the last hour before the test.
    - Don't talk to anyone about the test just before it; anxiety is contagious.

Test-Taking

- Read the directions carefully
    - This may be obvious, but it will help you avoid careless errors.
    - If there is time, quickly, look through the test for an overview.
    - Note key terms, jot down brief notes.
- Answer questions in a strategic order
    - First, easy questions-to build confidence, score points, and mentally orient yourself to vocabulary, concepts, and your studies (it may help you make associations with more difficult questions).
    - Then, difficult questions- With objective tests, first eliminate those answers you know to be wrong, or are likely to be wrong, don't seem to fit, or where two options are so similar that both must be incorrect.
- Review
    - Resist the urge to leave as soon as you have completed all the items.
    - Review your test to make sure that you have answered all questions, not mis-marked the answer sheet, or made some other simple mistake.
    - Do not "second-guess" yourself and change your original answers. Research has indicated that your first hunch is more likely to be correct.
    - You should only change answers to questions if you originally misread them or if you have encountered information elsewhere in the test that indicates with certainty that your first choice is incorrect.

Renal-Related Web Sites

- American Association of Kidney Patients (AAKP)
  www.aakp.org
- American Kidney Fund.
  www.akfinc.org
- American Nephrology Nurses' Association
  www.annanurse.org
- American Society for Artificial Internal Organs
  www.asaio.com
- American Society of Nephrology
  www.asn-online.org
- Centers for Disease Control and Prevention
  www.cdc.org
- American Society of Pediatric Nephrology
  www.aspneph.com
- Centers for Medicare & Medicaid Services (CMS)
  www.cms.gov
- Dialysis & Transplantation
  www.eneph.com
  A) Dialysisfinder
  www.dialysisfinder.com
  A) Dialysis Facility outcome Compare Home
  www.medicare.gov/Dialysis/Home.asp
- ESRD Networks
  www.esrdnetworks.org
- European Dialysis and Transplant Nurses Association/European Renal care Association (EDTNA/ERCA)
  www.edtna-erca.org
- Food and Drug Administration (FDA)
  www.fda.gov
- Hypertension, Dialysis, and Clinical Nephrology (HDCN)
  www.hdcn.com
- iKidney.com
  www.ikidney.com
- International Society of Nephrology
  www.isn-online.org
- International Society for Peritoneal Dialysis
  www.ispd.org

- Kidney Disease Outcomes Quality Initiative (K/DOQI)
  www.kidney.org/professionals/doqi/index.cfm
- Kidney & Urology Foundation of America
  www.kidneyurology.org
- Kidney School
  www.kidneyschool.org
- National Association of Nephrology Technicians/Technologists (NANT)
  www.dialysistech.net
- National Kidney Foundation
  www.kidney.org
- National Renal Administrators Association (NRAA)
  www.nraa.org
- National Transplant Assistance Fund (NTAF)
  www.transplantfund.org
- Nephron Information Center
  www.nephron.com
- National Institute of Diabetes and Digestive and Kidney Diseases (NIDDK)
  www.niddk.nih.gov
- PKD Foundation
  www.pkdcure.org
- Renal Physicians Association
  www.renalmd.org
- RENALNET
  www.renalnet.org
- Renal Support Network
  www.renalnetwork.org
- RenalWeb
  www.renalweb.com
- Transweb
  www.transweb.org
- United Network for Organ Sharing (UNOS)
  www.unos.org
- U.S. Renal Data System (USRDS)
  www.usrds.org

Renal Anatomy & Physiology

1. a b c d
2. a b c d
3. a b c d
4. a b c d
5. a b c d
6. a b c d
7. a b c d
8. a b c d
9. a b c d
10. a b c d
11. a b c d
12. a b c d
13. a b c d
14. a b c d
15. a b c d
16. a b c d
17. a b c d
18. a b c d
19. a b c d
20. a b c d
21. a b c d
22. a b c d
23. a b c d
24. a b c d
25. a b c d
26. a b c d
27. a b c d
28. a b c d
29. a b c d
30. a b c d
31. a b c d
32. a b c d
33. a b c d
34. a b c d
35. a b c d
36. a b c d
37. a b c d
38. a b c d
39. a b c d
40. a b c d
41. a b c d
42. a b c d
43. a b c d
44. a b c d
45. a b c d
46. a b c d
47. a b c d
48. a b c d
49. a b c d
50. a b c d
51. a b c d
52. a b c d
53. a b c d
54. a b c d
55. a b c d
56. a b c d
57. a b c d
58. a b c d
59. a b c d

Causes of Renal Failure

1. a b c d
2. a b c d
3. a b c d
4. a b c d
5. a b c d
6. a b c d
7. a b c d
8. a b c d
9. a b c d
10. a b c d
11. a b c d
12. a b c d
13. a b c d
14. a b c d
15. a b c d
16. a b c d
17. a b c d

Clinical Manifestations of Renal Failure

1. a b c d
2. a b c d
3. a b c d
4. a b c d
5. a b c d
6. a b c d
7. a b c d
8. a b c d

9. a b c d
10. a b c d
11. a b c d
12. a b c d
13. a b c d
14. a b c d
15. a b c d
16. a b c d
17. a b c d
18. a b c d
19. a b c d
20. a b c d
21. a b c d
22. a b c d
23. a b c d
24. a b c d
25. a b c d
26. a b c d
27. a b c d
28. a b c d
29. a b c d
30. a b c d
31. a b c d
32. a b c d
33. a b c d
34. a b c d
35. a b c d
36. a b c d
37. a b c d
38. a b c d
39. a b c d
40. a b c d
41. a b c d
42. a b c d
43. a b c d
44. a b c d
45. a b c d
46. a b c d
47. a b c d
48. a b c d
49. a b c d
50. a b c d
51. a b c d

Psychosocial

1. a b c d
2. a b c d
3. a b c d
4. a b c d
5. a b c d
6. a b c d
7. a b c d
8. a b c d
9. a b c d
10. a b c d
11. a b c d
12. a b c d
13. a b c d
14. a b c d
15. a b c d
16. a b c d
17. a b c d
18. a b c d
19. a b c d

Hemodialysis

1. a b c d
2. a b c d
3. a b c d
4. a b c d
5. a b c d
6. a b c d
7. a b c d
8. a b c d
9. a b c d
10. a b c d
11. a b c d
12. a b c d
13. a b c d
14. a b c d
15. a b c d
16. a b c d
17. a b c d
18. a b c d
19. a b c d
20. a b c d
21. a b c d
22. a b c d
23. a b c d
24. a b c d
25. a b c d
26. a b c d
27. a b c d

28. a b c d
29. a b c d
30. a b c d
31. a b c d
32. a b c d
33. a b c d
34. a b c d
35. a b c d
36. a b c d
37. a b c d
38. a b c d
39. a b c d
40. a b c d
41. a b c d
42. a b c d
43. a b c d
44. a b c d
45. a b c d
46. a b c d
47. a b c d
48. a b c d
49. a b c d
50. a b c d
51. a b c d
52. a b c d
53. a b c d
54. a b c d
55. a b c d
56. a b c d
57. a b c d
58. a b c d
59. a b c d
60. a b c d
61. a b c d
62. a b c d
63. a b c d
64. a b c d
65. a b c d
66. a b c d
67. a b c d
68. a b c d
69. a b c d
70. a b c d
71. a b c d
72. a b c d
73. a b c d
74. a b c d
75. a b c d
76. a b c d
77. a b c d
78. a b c d
79. a b c d
80. a b c d
81. a b c d
82. a b c d
83. a b c d
84. a b c d
85. a b c d
86. a b c d
87. a b c d
88. a b c d
89. a b c d
90. a b c d
91. a b c d
92. a b c d
93. a b c d
94. a b c d
95. a b c d
96. a b c d
97. a b c d
98. a b c d
99. a b c d
100. a b c d
101. a b c d
102. a b c d
103. a b c d
104. a b c d
105. a b c d
106. a b c d
107. a b c d
108. a b c d
109. a b c d
110. a b c d
111. a b c d
112. a b c d
113. a b c d
114. a b c d
115. a b c d
116. a b c d
117. a b c d
118. a b c d
119. a b c d
120. a b c d

121. a b c d
122. a b c d
123. a b c d
124. a b c d
125. a b c d
126. a b c d
127. a b c d
128. a b c d
129. a b c d
130. a b c d
131. a b c d
132. a b c d
133. a b c d
134. a b c d
135. a b c d
136. a b c d
137. a b c d
138. a b c d
139. a b c d
140. a b c d
141. a b c d
142. a b c d
143. a b c d
144. a b c d
145. a b c d
146. a b c d
147. a b c d
148. a b c d
149. a b c d
150. a b c d
151. a b c d
152. a b c d
153. a b c d
154. a b c d
155. a b c d
156. a b c d
157. a b c d
158. a b c d
159. a b c d
160. a b c d
161. a b c d
162. a b c d
163. a b c d
164. a b c d
165. a b c d
166. a b c d
167. a b c d
168. a b c d
169. a b c d
170. a b c d
171. a b c d
172. a b c d
173. a b c d
174. a b c d
175. a b c d
176. a b c d
177. a b c d
178. a b c d
179. a b c d
180. a b c d
181. a b c d
182. a b c d
183. a b c d
184. a b c d
185. a b c d
186. a b c d
187. a b c d
188. a b c d
189. a b c d
190. a b c d
191. a b c d
192. a b c d
193. a b c d
194. a b c d
195. a b c d
196. a b c d
197. a b c d
198. a b c d
199. a b c d
200. a b c d
201. a b c d
202. a b c d
203. a b c d
204. a b c d

Vascular Access

1. a b c d
2. a b c d
3. a b c d
4. a b c d
5. a b c d
6. a b c d
7. a b c d

8. a b c d
9. a b c d
10. a b c d
11. a b c d
12. a b c d
13. a b c d
14. a b c d
15. a b c d
16. a b c d
17. a b c d
18. a b c d
19. a b c d
20. a b c d
21. a b c d
22. a b c d
23. a b c d
24. a b c d
25. a b c d
26. a b c d
27. a b c d
28. a b c d
29. a b c d
30. a b c d
31. a b c d
32. a b c d
33. a b c d
34. a b c d
35. a b c d
36. a b c d
37. a b c d
38. a b c d
39. a b c d
40. a b c d
41. a b c d
42. a b c d
43. a b c d
44. a b c d
45. a b c d
46. a b c d
47. a b c d
48. a b c d
49. a b c d
50. a b c d
51. a b c d
52. a b c d
53. a b c d
54. a b c d
55. a b c d
56. a b c d
57. a b c d
58. a b c d
59. a b c d
60. a b c d
61. a b c d
62. a b c d
63. a b c d
64. a b c d
65. a b c d
66. a b c d
67. a b c d
68. a b c d
69. a b c d
70. a b c d
71. a b c d
72. a b c d
73. a b c d
74. a b c d
75. a b c d
76. a b c d
77. a b c d
78. a b c d
79. a b c d
80. a b c d
81. a b c d
82. a b c d
83. a b c d
84. a b c d
85. a b c d
86. a b c d
87. a b c d
88. a b c d
89. a b c d

Dialyzer Reprocessing

1. a b c d
2. a b c d
3. a b c d
4. a b c d
5. a b c d
6. a b c d
7. a b c d
8. a b c d

9. a b c d
10. a b c d
11. a b c d
12. a b c d
13. a b c d
14. a b c d
15. a b c d
16. a b c d
17. a b c d
18. a b c d
19. a b c d
20. a b c d
21. a b c d
22. a b c d
23. a b c d
24. a b c d
25. a b c d
26. a b c d
27. a b c d
28. a b c d
29. a b c d
30. a b c d
31. a b c d
32. a b c d
33. a b c d
34. a b c d
35. a b c d
36. a b c d
37. a b c d
38. a b c d
39. a b c d
40. a b c d
41. a b c d
42. a b c d
43. a b c d
44. a b c d
45. a b c d
46. a b c d
47. a b c d
48. a b c d
49. a b c d
50. a b c d
51. a b c d
52. a b c d
53. a b c d
54. a b c d
55. a b c d
56. a b c d
57. a b c d
58. a b c d
59. a b c d
60. a b c d
61. a b c d

Water Treatment

1. a b c d
2. a b c d
3. a b c d
4. a b c d
5. a b c d
6. a b c d
7. a b c d
8. a b c d
9. a b c d
10. a b c d
11. a b c d
12. a b c d
13. a b c d
14. a b c d
15. a b c d
16. a b c d
17. a b c d
18. a b c d
19. a b c d
20. a b c d
21. a b c d
22. a b c d
23. a b c d
24. a b c d
25. a b c d
26. a b c d
27. a b c d
28. a b c d
29. a b c d
30. a b c d
31. a b c d
32. a b c d
33. a b c d
34. a b c d
35. a b c d
36. a b c d
37. a b c d
38. a b c d

39. a b c d
40. a b c d
41. a b c d
42. a b c d
43. a b c d
44. a b c d
45. a b c d
46. a b c d
47. a b c d
48. a b c d
49. a b c d
50. a b c d
51. a b c d
52. a b c d
53. a b c d
54. a b c d
55. a b c d
56. a b c d
57. a b c d
58. a b c d
59. a b c d
60. a b c d
61. a b c d
62. a b c d
63. a b c d
64. a b c d
65. a b c d
66. a b c d
67. a b c d
68. a b c d
69. a b c d
70. a b c d
71. a b c d
72. a b c d
73. a b c d
74. a b c d
75. a b c d
76. a b c d
77. a b c d
78. a b c d
79. a b c d
80. a b c d
81. a b c d
82. a b c d
83. a b c d
84. a b c d
85. a b c d
86. a b c d
87. a b c d
88. a b c d
89. a b c d
90. a b c d
91. a b c d
92. a b c d
93. a b c d

Biomedical/ Chemistry

1. a b c d
2. a b c d
3. a b c d
4. a b c d
5. a b c d
6. a b c d
7. a b c d
8. a b c d
9. a b c d
10. a b c d
11. a b c d
12. a b c d
13. a b c d
14. a b c d
15. a b c d
16. a b c d
17. a b c d
18. a b c d
19. a b c d
20. a b c d
21. a b c d
22. a b c d
23. a b c d
24. a b c d
25. a b c d
26. a b c d
27. a b c d
28. a b c d
29. a b c d
30. a b c d
31. a b c d
32. a b c d
33. a b c d
34. a b c d
35. a b c d

36. a b c d
37. a b c d
38. a b c d
39. a b c d
40. a b c d
41. a b c d
42. a b c d
43. a b c d
44. a b c d
45. a b c d
46. a b c d
47. a b c d
48. a b c d
49. a b c d
50. a b c d
51. a b c d
52. a b c d
53. a b c d
54. a b c d
55. a b c d
56. a b c d
57. a b c d
58. a b c d
59. a b c d
60. a b c d
61. a b c d
62. a b c d
63. a b c d
64. a b c d
65. a b c d
66. a b c d
67. a b c d
68. a b c d
69. a b c d
70. a b c d
71. a b c d
72. a b c d
73. a b c d
74. a b c d
75. a b c d
76. a b c d
77. a b c d
78. a b c d
79. a b c d
80. a b c d
81. a b c d
82. a b c d
83. a b c d
84. a b c d
85. a b c d
86. a b c d
87. a b c d
88. a b c d
89. a b c d
90. a b c d
91. a b c d
92. a b c d
93. a b c d
94. a b c d
95. a b c d
96. a b c d
97. a b c d
98. a b c d
99. a b c d
100. a b c d
101. a b c d
102. a b c d
103. a b c d
104. a b c d
105. a b c d
106. a b c d
107. a b c d
108. a b c d
109. a b c d
110. a b c d
111. a b c d
112. a b c d

Labs

1. a b c d
2. a b c d
3. a b c d
4. a b c d
5. a b c d
6. a b c d
7. a b c d
8. a b c d
9. a b c d
10. a b c d
11. a b c d
12. a b c d
13. a b c d
14. a b c d

15. a b c d
16. a b c d
17. a b c d
18. a b c d
19. a b c d
20. a b c d
21. a b c d
22. a b c d
23. a b c d
24. a b c d
25. a b c d
26. a b c d
27. a b c d

Medications/ Anticoagulants

1. a b c d
2. a b c d
3. a b c d
4. a b c d
5. a b c d
6. a b c d
7. a b c d
8. a b c d
9. a b c d
10. a b c d
11. a b c d
12. a b c d
13. a b c d
14. a b c d
15. a b c d
16. a b c d
17. a b c d
18. a b c d
19. a b c d
20. a b c d
21. a b c d
22. a b c d
23. a b c d
24. a b c d
25. a b c d
26. a b c d
27. a b c d
28. a b c d
29. a b c d
30. a b c d
31. a b c d
32. a b c d
33. a b c d
34. a b c d
35. a b c d
36. a b c d
37. a b c d
38. a b c d
39. a b c d
40. a b c d
41. a b c d
42. a b c d
43. a b c d
44. a b c d
45. a b c d
46. a b c d
47. a b c d
48. a b c d
49. a b c d
50. a b c d
51. a b c d
52. a b c d
53. a b c d
54. a b c d
55. a b c d
56. a b c d
57. a b c d
58. a b c d
59. a b c d
60. a b c d
61. a b c d
62. a b c d
63. a b c d
64. a b c d
65. a b c d
66. a b c d
67. a b c d
68. a b c d
69. a b c d
70. a b c d
71. a b c d
72. a b c d

Nutrition

1. a b c d
2. a b c d
3. a b c d

4. a b c d
5. a b c d
6. a b c d
7. a b c d
8. a b c d
9. a b c d
10. a b c d
11. a b c d
12. a b c d
13. a b c d
14. a b c d
15. a b c d
16. a b c d
17. a b c d
18. a b c d
19. a b c d
20. a b c d
21. a b c d
22. a b c d
23. a b c d
24. a b c d
25. a b c d
26. a b c d
27. a b c d
28. a b c d
29. a b c d

Infection Control

1. a b c d
2. a b c d
3. a b c d

4. a b c d
5. a b c d
6. a b c d
7. a b c d
8. a b c d
9. a b c d
10. a b c d
11. a b c d
12. a b c d
13. a b c d
14. a b c d
15. a b c d
16. a b c d
17. a b c d
18. a b c d
19. a b c d
20. a b c d
21. a b c d
22. a b c d
23. a b c d
24. a b c d
25. a b c d
26. a b c d
27. a b c d
28. a b c d
29. a b c d
30. a b c d
31. a b c d
32. a b c d
33. a b c d
34. a b c d
35. a b c d
36. a b c d
37. a b c d
38. a b c d
39. a b c d
40. a b c d
41. a b c d
42. a b c d
43. a b c d
44. a b c d
45. a b c d
46. a b c d
47. a b c d
48. a b c d
49. a b c d
50. a b c d
51. a b c d
52. a b c d
53. a b c d

Peritoneal Dialysis

1. a b c d
2. a b c d
3. a b c d
4. a b c d
5. a b c d
6. a b c d
7. a b c d
8. a b c d
9. a b c d
10. a b c d

11. a b c d
12. a b c d
13. a b c d
14. a b c d
15. a b c d
16. a b c d
17. a b c d
18. a b c d
19. a b c d
20. a b c d
21. a b c d
22. a b c d
23. a b c d
24. a b c d
25. a b c d
26. a b c d
27. a b c d
28. a b c d
29. a b c d
30. a b c d
31. a b c d
32. a b c d
33. a b c d
34. a b c d
35. a b c d
36. a b c d
37. a b c d
38. a b c d

Other Treatment Modalities

1. a b c d
2. a b c d
3. a b c d
4. a b c d
5. a b c d
6. a b c d
7. a b c d
8. a b c d
9. a b c d
10. a b c d
11. a b c d
12. a b c d
13. a b c d

Miscellaneous

1. a b c d
2. a b c d
3. a b c d
4. a b c d
5. a b c d
6. a b c d
7. a b c d
8. a b c d
9. a b c d
10. a b c d
11. a b c d
12. a b c d
13. a b c d
14. a b c d
15. a b c d
16. a b c d
17. a b c d
18. a b c d
19. a b c d
20. a b c d
21. a b c d
22. a b c d
23. a b c d
24. a b c d
25. a b c d
26. a b c d
27. a b c d
28. a b c d
29. a b c d
30. a b c d
31. a b c d
32. a b c d
33. a b c d
34. a b c d
35. a b c d
36. a b c d
37. a b c d
38. a b c d
39. a b c d
40. a b c d
41. a b c d
42. a b c d
43. a b c d
44. a b c d

45. **a b c d**
46. **a b c d**
47. **a b c d**
48. **a b c d**
49. **a b c d**
50. **a b c d**
51. **a b c d**
52. **a b c d**
53. **a b c d**
54. **a b c d**
55. **a b c d**

Case Studies

1. **a b c d**
2. **a b c d**
3. **a b c d**
4. **a b c d**
5. **a b c d**
6. **a b c d**
7. **a b c d**
8. **a b c d**
9. **a b c d**
10. **a b c d**
11. **a b c d**
12. **a b c d**
13. **a b c d**
14. **a b c d**
15. **a b c d**
16. **a b c d**
17. **a b c d**
18. **a b c d**
19. **a b c d**
20. **a b c d**
21. **a b c d**
22. **a b c d**
23. **a b c d**
24. **a b c d**
25. **a b c d**
26. **a b c d**
27. **a b c d**
28. **a b c d**

Glossary

AAKP: American Association of Kidney Patients.

AAMI: Association for the Advancement of Medical Instrumentation. This organization sets the standards for dialysis machines and recommended practices for dialysis machines, reuse of dialyzers, electrical safety, monitoring and culturing of machines and water systems, cleaning of machines, quality of water used for dialysis and methodology for bacteriology and culturing samples.

Abscess: An infection under the skin that looks like a blister or pimple filled with fluid or pus.

ABG: Arterial blood gas; normal-- PCO_2: 25-40 mEq, HCO_3: 22-26 mEq, O_2: 95-100%.

Absolute: When referring to filters is used in reference to the micron rating of filters, indicating that a specified size will be trapped within or on the filter.

Absorption: The taking up of a substance into the physical structure of a liquid or solid by physical or chemical means, without a chemical reaction.

Access: A method to provide sufficient blood flow for hemodialysis through the surgical connection of a vein to an artery or the insertion of synthetic tubing into a central vein.

Acid: A substance that releases hydrogen ions when dissolved in water.

Acidity: An expression of the concentration of hydrogen ions present in a solution.

Acidosis: A condition in the body that results from the accumulation of acid or loss of alkali. In dialysis, a patient's condition prior to having a treatment will be described as metabolic acidosis due to the loss of bicarbonate.

Activated carbon: Carbon activated by high temperature steam or carbon dioxide to form a material of high adsorptive capacity. It is used primarily for the removal of chlorine and chloramine or dissolved organic matter in the water treatment system.

Acute: To indicate something is of short duration or sudden onset or to indicate a high degree of severity.

ADH: Antidiuretic hormone; produced and released by pituitary gland in response to low circulating volume.

Adsorption: The physical process by which molecules, colloids, or particles adhere to a surface without a chemical reaction (similar to how a magnet attracts and holds iron filings). The attachment of a substance to the surface of another. In dialysis, ß2 microglobulin is adsorbed onto the membrane of a dialyzer during a treatment.

Advance Directives: Documents that outline the patient's wishes for treatment or no treatment, in case the patient becomes too ill to make such decisions at a later date.

Afferent: Toward an organ.

AGE: Advanced glycosylated end-products. AGEs are products of glucose metabolism and are linked to diabetes mellitus. The higher a person's blood sugar, the more AGEs they tend to have in the blood. These AGEs, cause much damage to the body, including kidney damage, plaque buildup on the arteries, stroke, high blood pressure, damage to the lens of the eye, blindness, and may accelerate the aging process in general.

Agglomerate: The process of bringing together smaller divisions into a larger mass.

AIDS: Acquired Immune Deficiency Syndrome

Air embolus: An air bubble carried by the bloodstream to a vessel small enough to be blocked by the bubble.

Albumin: A protein found in many body tissues. It disperses in water as a colloid and is an important fraction of blood plasma. A large protein of 68000 daltons found throughout the body. It is used to increase blood osmolality during dialysis to improve the movement of interstitial fluid into the blood stream.

Aldosterone: A hormone secreted by the adrenal cortex to regulate electrolytes and water balance. It causes the kidneys to retain sodium and excrete potassium.

Alkalosis: A condition in the body that results from the accumulation of base (generally bicarbonate) or loss of acid.

Alum: Aluminum sulfate; widely used for flocculation of contaminants in municipal water treatment, typically for the removal of color and turbidity.

Amino Acids: Building blocks of protein.

Ampere: The unit of measure for electron flowrate.
One ampere of current is a flowrate of 6.24×10^{18} electrons per second past a point in a circuit. The ampere is named after the French physicist and mathematician Andre Marie Ampere (1775-1836). One thousandth of an ampere is called a milliampere (mA). Fuses are sized in amperes, which sets the limit of electrical current that they can tolerate before blowing.

Amyloidosis: Build-up of a protein called beta-2 microglobulin (B_2) in the soft tissues, bones and joints. In dialysis patients, these deposits can cause arthritis-like joint pain and/or bone pain. An accumulation of amyloid in organs or tissue can cause loss of function.

Analgesic: A medication that relieves pain.

Anaphylaxis: An immediate, severe reaction to a substance to which an individual is allergic. The reaction may include hives, itching, or wheezing. It may progress to include hypotension and spasms of the breathing passages and death.

Anastomose: To unite surgically; in dialysis a surgically created connection between an artery and a vein to form a fistula suitable for repeated access.

Anemia: A shortage of oxygen-carrying red blood cells. It causes severe fatigue, heart problems, trouble concentrating, reduced immune function and other problems. The condition of having a reduced number of red blood cells. A healthy individual will usually have a hematocrit of at least 40%, whereas a typical renal patient may be in the 20's due to loss of production of erythropoietin by the kidneys.

Anesthetic: A drug that numbs the body to reduce pain.

Aneurysm: A blood filled sac formed by stretching and thinning of the wall of an artery.

Angina: Chest pain usually caused by a lack of oxygen to the muscles of the heart.

Angioplasty: A procedure to open a narrowed blood vessel (stenosis).

Angiotensin I: A polypeptide in the blood that is formed by the action of renin on angiotensinogen in the blood plasma. It is converted to Angiotensin II, a powerful blood pressure increasing hormone, by an enzyme found primarily in the lungs.

Angiotensin II: A hormone that raises blood pressure and reduces fluid loss in the kidneys by restricting flow.

Angstrom: Unit of length equaling 10^{-10} meters.

Anion: Ion carrying a negative electric charge.

ANNA: American Nephrology Nurses Association.

Anorexia: Lack of appetite for food.

Antacids: Medication used to bind phosphorus in the intestines to supplement the loss of this function by the kidneys. Without their use, bone disease will develop in the ESRD patient.

Antegrade: Forward moving. In dialysis, it means in the direction of blood flow.

Antibody: Protein produced in the body in response to invasion by a foreign substance.

Anticoagulant: Medication or chemical that prevents clotting of the blood.

Anticoagulation: Method to prevent coagulation of blood by inhibition of coagulation factors e.g. factor Xa and 11a (thrombin). Commonly used during extracorporeal circulation to prevent coagulation in the device/circuit.

Antigen: Substance introduced into body that induces a state of sensitivity and/or immune responsiveness.

Antiseptics: Products that slow or stop the growth of bacteria viruses. Used to kill micro-organisms to prevent infection and the spread of disease.

Anuria: Urine output of less than 50 mL/day. Normal urine output is approximately 1500mL/day.

Apical pulse: Felt on the chest wall directly over the heart.

Apnea: Temporary period when breathing stops.

ARF: Acute Renal Failure.

Arrhythmia: Any variation in the normal rhythm of the heartbeat.

Arterial Pressure: In hemodialysis, arterial pressure is measured between the arterial needle and dialyzer. Pre-pump pressure is measured from the patient's access to the blood pump. Post-pump arterial pressure is measured after the blood pump and before the dialyzer.

Arteriole: A small artery.

Arteriovenous Fistula (AVF): A fistula is a surgical connection between an artery and vein beneath the skin in the upper or lower arm to provide access to the blood.

Artery: A blood vessel that carries blood away from the heart at high pressure. Arteries deliver oxygenated blood to every part of the body.

Artificial: Man-made, often an imitation of something found in nature.

Ascites: Build-up of fluid in the abdomen caused by liver damage, heart failure, malnutrition or infection.

Asepsis: Absence of pathogens.

Aseptic: Free of bacterial or infectious organisms.

Aseptic Technique: Series of steps used to maintain a germ free environment.

Aspirate: Remove something by suction or negative pressure.

Atherosclerosis: Hardening of the arteries caused by a buildup of yellowish plaque in the walls of the arteries.

Atom: The smallest unit of an element, such as hydrogen or oxygen.

Atony: Lack of normal tone or strength.

ATP: Adenosine triphosphate; compound which extracts or supplies energy in cellular reactions.

ATN: Acute tubular necrosis.

Auscultate: Listen with a stethoscope.

Azotemia: Retention of nitrogenous wastes, such as urea and creatinine in blood and body fluid.

Backfiltration: Movement of dialysate across the dialyzer membrane and into the patient's blood. It can be caused by a change in the pressure or concentration gradient between the dialysate and the blood.

Backwash: Reversal of a solution's flow through a filtration system in which beds of filter or ion exchange media are subjected to flow in the opposite direction from the service flow to loosen the bed and flush out suspended matter.

Bacteremia: A presence of bacteria in the bloodstream.

Bacteria: Unicellular organisms that typically reproduce by cell division and are prevalent everywhere.

Base: A substance that releases hydroxyl ions when dissolved in water.

Bicarbonate: A buffer used by the body to neutralize acids that form when the body breaks down protein and other foods. Bicarbonate is used in dialysate to help restore levels of bicarbonate in the body.

Bilirubin: A byproduct of RBC breakdown; secreted in to the bile to be used again.

Biocompatible: Not causing change or reaction in living tissue. The ability of a material, device or system to perform without a clinically significant host response. The dictionary definition of biocompatibility is - "a quality of being mutually tolerant with life" and is derived from the Greek (bios-meaning course of life) and the Latin (compatibilities- meaning mutually tolerant of). A biocompatible membrane would not damage blood cells, cause clotting or release pyrogenic matter. Protein adsorption to membrane surface occurs instantaneously and leads to platelet adhesion and activation of the coagulation and complement pathways. Transient leukopenia at the onset of dialysis is a common occurrence. The ability of a membrane to adsorb proteins to the fiber wall is the primary mechanism in membrane biocompatibility. These adsorbed proteins, called a secondary membrane, effectively isolate the blood which is then no longer exposed to the foreign substances. This is the reason that reused dialyzers are more biocompatible than are new ones. Synthetic membranes are more adsorptive than cellulose membranes due to their hydrophobic nature. Described as membrane biocompatibility, interactions between soluble and cellular components of the blood and selected dialysis membrane materials result in perturbations in the complement cascade and granulocyte number and function. As a consequence of these membrane - associated immunologic abnormalities, ESRD patients may be at increased risk of malnutrition, infection, hospitalization and death.

Biodegradation: Processes by which living systems render chemicals less noxious to the environment.

B_2M: Beta-2 microglobulin.

Blood leak: Occurs when the semipermeable membrane of the dialyzer tears, letting blood leak into the dialysate.

Blood pump: Pushes the patient's blood through the extracorporeal circuit at a fixed rate of speed.

Blood tubing: Carries blood from the patient's vascular access through the arterial needle to the dialyzer and back to the patient through the venous needle.

Bolus: A dose of medication given all at once intravenously.

BONENT: Board of Nephrology Examiners in Nursing and Technology.

Bowman's Capsule: The capsule that houses the glomerulus where fluid is initially removed from the blood to begin the filtration process in the kidney. Named after Sir William Bowman, an English physician (1816-1892).

Brachial Pulse: Pulse felt in the crease of the elbow at the brachial artery.

Brachiocephalic Fistula: Type of AV fistula in the upper arm. Created by surgically joining the brachial artery and the cephalic vein.

Bradycardia: Slow heart rate.

Brine: A solution of sodium chloride used for regenerating water softeners.

Bruit: A buzzing or swooshing sound caused by high pressure flow of blood through an arterio-venous fistula or graft. A high-pitched bruit may mean there is a stenosis in the access. should be low pitch

Buffer: A chemical that causes a solution to resist changes in pH. Acetate and bicarbonate are the two buffers used in dialysis to maintain the pH of dialysate.

BUN: Blood Urea Nitrogen; by-product of protein metabolism.

Buttonhole Technique: (constant site) A cannulation technique, where dialysis needles are placed in a fistula into the same holes at the same angle.

Bypass: Safety feature of the hemodialysis delivery system that prevents the flow of dialysate to the dialyzer and sends it to the drain when the dialysate fails to meet the safety condition of conductivity, temperature or pH. Bypass prevents unsafe dialysate from reaching the patient and causing harm.

Cachexia: General ill health and malnutrition. Wasting.

CAD: Coronary Artery Disease; narrowing of heart vessels due to deposits of cholesterol and lipids.

Calcitriol-Calcijex: Calcitriol is used to treat hypocalcemia in patients receiving chronic dialysis. It is the active form of Vitamin D_3. It increases calcium levels and has been shown to reduce elevated PTH levels, preventing secondary hyperparathyroidism and improving renal osteodystrophy. Calcitriol is available in an intravenous (IV) form or oral form (Rocaltrol).

stenosis - narrowing of the vessel

Calcium: A positively charged ion (cation). In the body, calcium is an electrolyte needed for nerve and muscle function and to form normal bone.

Calibrate: Adjust or accurately set a measuring device by comparison with a known standard.

Cannulate: To place dialysis needles into a fistula or graft.

Capillaries: The body's smallest blood vessels, where oxygenated blood crosses from arteries to veins.

CAPD: Continuous Ambulatory Peritoneal Dialysis.

Carbon Tank: Water treatment devices that contain granular, activated carbon that absorbs low molecular weight particles from water.

Cardiac arrest: Occurs when the heart stops beating.

Cardiac arrhythmia: An irregular heartbeat.

Cardiac output: Amount of blood passing through the heart in a certain period of time.

Catabolism: Any destructive metabolic process by which organisms convert substances into excreted compounds.

Catheter: A hollow plastic tube for withdrawing or introducing fluid into a cavity or passage of the body.

Cation: Ion carrying a positive charge.

CAVH: Continuous Arterio-Venous Hemofiltration.

CAVHD: Continuous Arterio-Venous Hemodialysis.

CAVU: Continuous Arterio-Venous Ultrafiltration.

CBNT: Certified in Biomedical Nephrology Technology.

CCHT: Certified Clinical Hemodialysis Technician.

CCNT: Certified in Clinical Nephrology Technology.

CCPD: Continuous Cycling Peritoneal Dialysis.

CDC: Centers for Disease Control & Prevention.

Cellulase: An enzyme which causes the decomposition of cellulose.

Cellulose: A fiber that forms the cell walls of plants.

Central Venous Stenosis: Narrowing of a central vein.

Channeling: The flow of water or other solution in a limited number of passages in a filter or ion exchange bed instead of distributed flow through all the passages in the bed.

CHF: Congestive Heart Failure; inability of the heart to pump out the blood returned to it.

Chloramine: Chemical compound (combined chlorine) containing chlorine attached to nitrogen, which retains its bactericidal qualities for a longer time than does free chlorine. Chloramine is formed when chlorine and ammonia are mixed in water. Chloramine is an oxidant, a substance that destroys microorganisms by breaking down their cell walls. It is used to kill bacteria in water and can be removed by activated carbon. Chloramine levels can be measured indirectly by measuring both total chlorine and free chlorine and calculating the difference.

Chloride: A salt concentrate needed in dialysate and in the human body.

Chronic: Long-term.

Chronic Kidney Disease (CKD): A long, usually slow, progressive loss of nephrons (results in loss of kidney function).

CHT: Certified Hemodialysis Technician.

Clearance: Mathematical expression of the rate at which a given substance is removed from a solution. In dialysis, it is a measure of net flux across the hemodialyzer membrane expressed as the number of milliliters of blood completely cleared of a solute. For the purpose of this recommended practice, clearance includes clearance due to ultrafiltration. Amount of blood (in mLs) that is completely cleared of a solute in one minute of dialysis at a given blood and dialysate flow rate.

$$C_x = [(A_x - V_x)/A_x]Q_B$$

Where: A_x = concentration at inlet (arterial)
V_x = concentration at outlet (venous)
Q_B = blood flowrate (in mL/min)

CMS: Centers for Medicare and Medicaid Services.

Coagulant: Chemical added in water to cause the agglomeration of finely divided particles into large particles which can be removed by settling or filtration; promoting, accelerating, or making possible coagulation of blood.

Coalesce: To merge or unite into a single body or mass. Coalescing is created by a filter in order to bring small microbubbles in solution together to form larger bubbles that can then be removed from the water. Coalescing is a part of the overall function of deaeration.

Coefficient of ultrafiltration (KuF): The fixed amount of fluid a dialyzer will remove from the patient's blood per hour at a specified pressure. KuF is also called the ultrafiltration factor (UFF) or ultrafiltration rate (UFR). It is expressed as millimeters (mL) per hour (hr) of water removed for each millimeter (mm) of mercury (Hg) of transmembrane pressure (TMP) or mL/hr/mmHg TMP.

Colloid: A very finely divided substance, larger than a molecule that spreads throughout a liquid as tiny particles. It exerts an osmotic effect proportionate to its concentration (serum albumin is a colloid).

Colony forming unit (CFU): The number of CFU's in a water or dialysate sample is a measure of the number of living bacteria.

Comorbid: A coexisting illness or disease process not directly related to the primary disorder.

Complement activation: The activation of the complement cascade by either the classical or alternate pathway inducing leukopenia and other symptomatology. In biocompatibility, it is activation as a result of blood/material or blood/device interaction or heparin/protamine interaction.

Compliance: The capability of stretching or yielding to pressure. Generally associated with dialyzer membranes in flat plate dialyzers whose blood volumes vary with the blood pressure.

Compound: A distinctive substance formed by the chemical union of two or more elements in definite proportion by weight.

Concentrate: One of two salt solutions (acid & bicarbonate) that are mixed together to form dialysate.

Concentration: Amount of solute dissolved in a measure of fluid.

Concurrent: Dialysis fluid and blood flow are in the same direction.

Conductivity: The ability of an aqueous solution to carry electric current depends on the presence of ions in the solution. It depends on temperature and total concentration of ionized substances in solution. The measurement of water's electrical conductivity can provide an assessment of total ionic concentration. Highly purified water is a poor conductor. Conductivity is expressed in units of Siemen/cm.

Conductivity Cell: A device designed to monitor the ability of a solution to conduct electricity. Cells used specifically for measuring the conductance of dialysate solutions are specifically designed to perform this function. The utilization of a temperature monitoring device in conjunction with a cell is necessary if total ion concentration is to be monitored.

Congestive Heart Failure (CHF): The inability of the heart to pump effectively due to excessive fluid in the bloodstream.

Convection: This process (or solvent drag) occurs when solutes are carried along by a flowing solvent as a result of pressure gradient. In dialysis, any molecules small enough to pass through the filter can be removed from the blood by convection.

Cortex: An outer layer of an organ. The renal cortex is the location of Bowman's Capsule where each nephron begins with the filtering of fluid from the blood across the glomeruli.

Coumadin: An oral medication (Sodium Warfarin) used to help prevent blood clotting in a patient's access.

Countercurrent: Direction of flow of dialysis fluid and of blood are 180° opposite one another.

CPR: Cardio-Pulmonary Resuscitation. CPR is a medical procedure used to restart a patient's heart and breathing when a patient suffers heart failure.

CQI: Continuous Quality Improvement.

Creatinine: One of the nitrogenous waste products of normal muscle metabolism.

Creatinine clearance: A test that measures how efficiently the kidneys remove creatinine from the blood.

Crenation: Crenation occurs if the blood cells are exposed to a solution more concentrated than blood and the blood will appear dark and red.

CROWNWEB: Consolidated Renal Operations in a Web enabled Network. A web based software application that all dialysis facilities will be required to use by end of 2009

CRF: Chronic Renal Failure; irreversible, gradual onset of renal failure.

CRRT: Continuous Renal Replacement Therapy.

CTS: Carpal Tunnel Syndrome; a painful condition of the hand caused by compression of the median nerve in the hand. The problem can occur in longer term hemodialysis due to deposition of amyloid at the carpal tunnel sheath. It causes pain, numbness and tingling of the thumb.

CVHD: Continuous Veno-Venous HemoDiafiltration.

CVVH: Continuous Veno-Venous Hemofiltration.

Cyanosis: A bluish discoloration of the skin due to reduced hemoglobin in the bloodstream. This condition is caused by a deficiency of oxygen in the blood.

Cytokines: A family of polypeptides with a molecular weight of 10,000-25,000 Daltons which is produced by different cells following or as part of a response to immunological interactions or inflammatory stimuli; they regulate the intensity and duration of immune responses.

Dalton: A unit of mass 1/12 the mass of carbon. Named after John Dalton, founder of atomic theory.

Deionization: Removal of solute ions from a solution; in a water treatment process, removes all the electrolytes from the water by exchange with other ions associated with fixed charges on a resin. Positively charged ions are removed by a cation exchange resin in exchange for a chemically equivalent amount of hydrogen ions. Negatively charged ions are removed by anion exchange resin for a chemically equivalent amount of hydroxide ions. Deionization typically does not remove organic compounds, viruses, or bacteria.

Delta P: A commonly used term denoting the pressure drop across a filter.

Dementia: Progressive decline in cognitive function due to damage or disease in the brain.

Dextrose: A simple sugar, readily used by body cells for metabolism.

Dialysance: The capability of a dialyzer to clear a given solute. Represents the net rate of exchange of a substance between blood and dialysate/minute/unit of blood-dialysate concentration gradient

Dialysate: An electrolyte solution used in dialysis to remove excess fluids and metabolic end products from the blood by a diffusion gradient across the membrane, while maintaining physiological concentrations of electrolytes and sometimes glucose.

Dialysis: The separation of substances in a solution by means of their unequal rate of diffusion through semipermeable membranes.

Dialysis dementia: A severe, often fatal encephalopathy which has been attributed to accumulation in the brain of aluminum from dialysate prepared with inadequately purified water.

Diastole: The period of time when the heart relaxes to allow filling to occur. *See systole.*

Diffusion: A movement of particles from an area of a higher concentration to a lower concentration; in dialysis, metabolic waste products are transported from the blood into the dialysate by this process.

Dilute: Thin out or weaken. A solution is made less concentrated (diluted) by the addition of more solvent.

Disequilibrium syndrome: A condition caused by too rapid hemodialysis, believed to be due to brain swelling. In its mildest form, the syndrome is limited to restlessness, headache, nausea and vomiting. During dialysis, blood solute is cleared faster than brain solute and an osmolar shift takes place between blood and brain. This is more likely to occur when patients with advanced states of uremia are dialyzed for an excessive period of time during their first treatment. (With hemodialysis, the concentration of BUN is often reduced more rapidly than the urea nitrogen in the cerebrospinal fluid and brain tissue. This is due to the relatively slow transport of urea across the blood-brain barrier via the vertebral spinal fluid (CSF). Osmolality in the CSF falls more slowly than the brain; pressure increases during dialysis and cerebral edema occurs).

Disinfectant: A chemical or gas which is able to kill most bacteria, but normally not spores.

Distal: Away from a point of reference such as the feet, hands, head or kidney. The opposite of proximal.

Diuresis: An increased output of urine.

Diverticulum: A pocket or sac formed in the wall of a tube or vessel.

DNR: Do Not Resuscitate.

Doxicaliciferol (Hectorol): A synthetic vitamin D analog used to suppress PTH and manage secondary hyperparathyroidism available in intravenous or oral form. Hypercalcemia, hyperphosphatemia & over suppression of the parathyroid gland are possible adverse effects associated with the use of this medication. The dosing is based on PTH levels along with the monitoring of the serum calcium and phosphorous.

DRG: Diagnosis Related Groups, The DRG system categorizes the entire range of reasons people are hospitalized into about 600 groups to determine how much the hospital will be paid by the insurance company.

Dry weight: Weight of dialysis patient when the blood pressure is normal and all excess fluid has been removed.

Dwell time: The length of time dialysis solution remains in the abdominal cavity during peritoneal dialysis before it is drained and replaced with fresh dialysate.

Dyspnea: Difficulty breathing or shortness of breath.

Dyspraxia: Partial loss of the ability to perform coordinated acts.

E. Coli: *Escherichia coli*, a bacteria associated with animal and human waste.

EBCT: Empty Bed Contact Time; a measurement of how much contact occurs between particles, such as activated carbon and water as the water flows through a bed of the carbon particles. AAMI recommendation: minimum 10 minutes.

$$\text{EBCT (minutes)} = (\text{volume of carbon } (ft^3) \times 7.48)/\text{Max. } H_2O \text{ Flow (gal/min)}$$

Ecchymosis: A small hemorrhagic spot in the skin that forms a nonelevated, rounded or irregular blue or purplish patch. A bruise.

Edema: Collection of fluid in body tissue causing swelling, often soft and compressible. Also abnormal accumulation of fluid in intercellular spaces of the body.

Edematous: The presence of an excessive amount of fluid in the intracellular compartment resulting in the area appearing puffy, swollen and the skin stretched out.

EDW: Estimated Dry Weight.

Effluent: The outflow from something.

Electrolyte: Any ion or solution of ions, capable of transferring or exchanging electrons. In dialysis fluid, it is the charged ions resulting from dissociation of salts when dissolved in the water.

Embolus: A clot or clot fragment carried by the bloodstream to a smaller vessel resulting in its blockage.

Empirical: Founded on practical experience but not proved scientifically.

Encephalopathy: Any gross dysfunction of the brain, temporary or permanent, that may result from anatomic damage, metabolic imbalance, or toxic agents; it is associated with long term use of hemodialysis and marked by speech disorders and constant muscle jerks progressing to dementia. Caused by high levels of aluminum accumulating in the body from either the water source used for dialysis or aluminum based phosphate binders used by the patient.

Endocarditis: Inflammation of the membrane that surrounds the heart as well as the supportive tissue bed on which it lies.

Endocrine: To secrete internally. An organ in the body that secretes a hormone into the blood stream or lymph glands to cause a specific action by another organ.

Endogenous: Originating within the body.

Endotoxin: A toxic substance produced and held within cell walls of bacteria until they die or are destroyed, where upon it may be released. The complex phospholipid-polysaccharide macromolecule which form an integral part of the cell wall of Gram-negative bacteria including the enterobacteria. Toxic substance (lipopolysaccharide) from gram negative bacteria that has a broad spectrum of biological activities, including pyrogenicity. There is an endotoxin limit by AAMI recommended for Reuse water of 2 EU/ml. Endotoxin is ONLY present in gram-negative bacteria with one exception: Listeria monocytogenes.

Equilibrium: State of balance between opposing forces.

Equivalent: The proper term is actually one gram equivalent weight. An equivalent of an element is the amount of that element which will combine or replace one gram of hydrogen. This amount is the same as the gram atomic weight of an element divided by its valence. Equivalent can also be defined as the amount of an element necessary to get 6.02252×10^{23} (Avogadro's Number) of positive or negative charges when the element is dissolved.

Erythrocyte: Red blood cell.

Erythropoiesis: Process of making red blood cells by the bone marrow.

Erythropoietin: A hormone, normally produced by the kidneys, that stimulates the bone marrow to produce red blood cells (erythropoiesis).

ESRD: End-Stage Renal Disease defined as <15% of normal GFR.

Ethylene oxide: A gas which is capable of sterilization by alkylating NH_2 and OH groups. Residual absorbed gas in dialyzer materials may result in hypersensitivity reactions.

EU: Endotoxin Units as used with Limulus Amebocyte Lysate (LAL) when testing for endotoxin. Standard unit of activity is based on a standard solution of Ecol. endotoxin.

Euvolemia: Normal intravascular volume.

Exchange: In peritoneal dialysis, the procedure of discarding the used dialysis solution and instilling fresh dialysis solution in the abdominal cavity.

Excretion: The act of discarding or eliminating. Urine is excreted from the kidney. The term can also be a direct referral to the excreted substance itself.

Exhaustion: The point where the resin is no longer capable of additional exchange.

Exogenous: Originating outside the organism.

Extracellular fluid: Fluid within the body, but outside the cells. This fluid can be subdivided into two types: intravascular and interstitial. It is the interstitial fluid that must move into the intravascular area if it is to be removed by the dialyzer.

Extracorporeal: Outside the body.

Febrile: Feeling feverish; relating to elevated body temperature.

Fecal: Excretion from the bowel.

Ferritin: The storage form of iron (normal >100 ng/ml).

Fiber Bundle Volume (FBV): The aggregate volume of patent hollow fibers contained within the blood compartment of a hollow fiber dialyzer.

Fibrin sheath: A collection of blood clotting fibers that build up in the outside of a catheter lumen.

Filtration: The passage of a liquid through a filter resulting from a pressure drop across the filter.

First use syndrome: A hypersensitivity reaction observed in patients treated with a new dialyzer. A symptom complex characterized by nervousness, chest pain, back pain, palpitations, pruritus and other usually mild symptoms, occurring about 15 minutes following the initiation of dialysis with a new dialyzer. Reuse of dialyzers is associated with a marked reduction of symptoms. The symptom complex may be associated with the induction of complement activation and leukopenia. The syndrome is defined by some authorities to include the anaphylactoid occurring immediately after the initiation of dialysis in some patients using dialyzers sterilized with ethylene oxide.

Fistula: An abnormal opening or passage. As related to dialysis, a surgical opening between an artery and vein to fill the vein with arterial blood. The force of the blood flow enlarges the vein and thickens the vein wall, allowing repeated punctures to obtain the large volume of blood needed for dialysis.

Flame Photometer: This instrument is used to measure sodium and potassium ion concentrations in various solutions including dialysate. The sample to be tested is placed in a flame and the intensity of the color produced is translated into ion concentrations expressed in milliequivalents per liter (mEq/L)). Typical results of dialysate solution for sodium would be 135-140 mEq/L, and for potassium 0.0-3.0 mEq/L.

Flocculation: The agglomeration of finely divided particles into larger groupings called floc, which then settle by gravity.

Flux rate: The rate per unit of area at which water passes through a semipermeable membrane per unit of time. Usually expressed in gallons per square foot per day (GFD).

Fouling: The deposition of insoluble materials, such as bacteria, colloids, oxides and water-borne debris, onto the surface of a reverse osmosis, ultrafiltration membrane, in a bed of filter media or ion exchanger. Fouling is associated with decreased flux rates and may also reduce the rejection rates of R.O. membrane.

Free Chlorine: Dissolved molecular chlorine.

Fribrin: White particles created during the early stages of the blood clotting process.

Gamma irradiation: Gamma rays are a form of high frequency, high energy radiation emitted from radioactive atomic nuclei. They kill all bacteria, spores, & viruses that they hit.

Germicides: A chemical which is able to kill microorganisms, but not necessarily all types.

GFD: Gallons per square Foot per Day.

GFR: Glomerular Filtration Rate; the volume of blood filtered by the glomerulus each minute in mL/min. (normal 125 mL/minute).

Globulin: A family of proteins found in serum and tissue of much larger molecular size than albumin. Certain serum globulins are involved in the immune response of the body and are called immunoglobulin (antibodies).

Glomerulonephritis: Inflammation of the capillary loops in the glomerulus.

Gradient: Rate of increase or decrease between two variables.

Grains per gallon: Unit of concentration equal to 17.1 milligrams per liter. 1 pound = 7,000 grains.

Gram Stain: A colorful stain used to visualize bacteria which are usually colorless and invisible to light microscopy; separates organisms into 2 groups: gram-positive and gram-negative. When the slide is studied microscopically, cells that absorb the crystal violet stain (a blue dye) and hold onto it will appear blue. These are called gram-positive organisms. However, if the crystal violet stain is washed off by alcohol, these cells will absorb the safranin (a red dye) and appear red. These are called gram-negative organisms. The different stains are the result of differences in the cell walls of gram-positive and gram-negative bacteria.

Gu: Net urea generation rate. It can be calculated from the change in body urea content from the end of one dialysis to the beginning of the next dialysis.

Half-life: The half-life is the amount of time it takes for the excretion or elimination process of the body to decrease the effect of the drug by one-half. The half-life of a drug is an important element in determining the appropriate dosage intervals necessary for a drug to maintain a therapeutic concentration level in the body.

Hardness: A measure of the ability of water to precipitate soaps made from fatty carboxylic acids. These soaps precipitate in the presence of calcium and/or magnesium ions. The hardness is used to describe the total concentration of calcium and magnesium, expressed as mg/L or gr/gal of calcium carbonate.

HCFA: Health Care Financing Administration.

Hematocrit: The percentage of RBC in whole blood.

Hematoma: A painful, discolored swelling over a puncture site usually caused by bleeding under the skin. Caused by blood leaking out of a vessel and into the surrounding tissue.

Hematuria: The presence of red blood cells or blood in the urine.

Hemoconcentration: Dehydration of blood.

Hemofiltration: Removal of water from the blood by ultrafiltration without dialysis.

Hemoglobin: Oxygen-carrying portion of the red blood cells. Respiratory protein of erythrocytes, that has the capacity to bind oxygen and carry it throughout the body.

Hemolysis: Breaking up of red blood cells resulting in release of hemoglobin into the surrounding fluid. Hemolysis may be caused by specific complement-fixing antibodies, toxins, various chemical agents, heating, etc.

Hemolytic anemia: An anemia resulting from the destruction of red blood cells.

Hemoperfusion: Removal of noxious substances by passing blood over a column of charcoal or special resin that has high binding capacity.

Hemostasis: The arrest of bleeding.

Heparin: An anticoagulant chemical which prevents platelet agglutination and thus prevents thrombus formation.

Hepatitis: Inflammation of the liver, usually from a viral infection, sometimes from toxic agents.

High level disinfection: Physical or germicidal procedure that results in inactivation of all vegetative microorganisms.

High-efficiency dialysis: Dialysis utilizing a greater than conventional dialyzer clearance; the coefficient factor of the membrane normally falls between 5 and 15.

High-flux dialysis: Uses a membrane permeable to a broad range of molecular weight solutes. The coefficient factor of the membranes is generally higher than 15 and need ultrafiltration control machines to do the dialysis procedure safely.

HIPAA: Health Insurance Portability and Accountability Act.

HIV: Human Immunodeficiency Virus.

Hg: Mercury

Homeostasis: The ability of the body to maintain normal body function through a series of negative feedback mechanisms. An example is when carbon dioxide builds up in the blood, breathing is increased to reduce the concentration back to normal.

Hydrophilic: A substance that blends or readily absorbs water (water-loving); a property of material characterized by low surface energy thereby attracting water molecules.

Hydrophobic: A substance which does not readily absorb water (water-rejecting); a property of material characterized by high surface energy resulting in the repulsion of water molecules and has thus an increased capacity to adsorb proteins and cells.

Hydrostatic pressure: The force exerted by a liquid on the walls of its container. It can be positive or negative. A positive pressure will try to push the walls out, whereas a negative pressure will try to suck the sides in.

Hyper: Prefix to indicate excessive or above the normal standard.

Hypercalcemia: An excess of calcium in the blood. Manifestations include fatigability, muscle weakness, depression, anorexia, nausea and constipation. Normal levels in blood are 9.0-10.5 mg/dL or 2.25-2.75 mmols/L.

Hyperglycemia: A higher than normal blood sugar level.

Hyperkalemia: Above normal levels of potassium in the blood. Normal levels are 3.5-5.0 mEq/L.

Hypernatremia: Above normal levels of sodium in the blood. The normal range is 135-145 mEq/L. This condition leads to excessive water being retained by the patient.

Hyperphosphatemia: Excessive amounts of phosphate in the blood. Normal blood levels are 2.6-4.6 mg/dL.

Hypersensitivity: Above normal sensitivity or allergy.

Hypertension: Above normal blood pressure. In the ESRD patient, this condition is often caused by the retention of excess fluid.

Hypertonic: A solution which contains a higher concentration of solutes than another solution.

Hypertonic Saline: A sodium chloride solution that contains 26 times more sodium chloride than normal saline. Normal saline is a 0.9% (9 grams/liter) solution of sodium chloride. Hypertonic saline is a 23.4% (234 grams/L) solution of sodium chloride.

Hypertrophy: The enlargement of an organ or a part of an organ due to an increase in the size of its cells.

Hypervolemia: The existence of excess water in the intravascular fluid compartment. This condition can result in hypertension or congestive heart failure.

Hypo: Prefix to indicate lower or less than the normal standard.

Hypocalcemia: A below normal level of calcium in the blood.

Hypokalemia: A below normal level of potassium in the blood. Normal levels in the blood are 3.5-5.0 mEq/L.

Hyponatremia: A below normal level of sodium in the blood. The normal range is 135-145 mEq/L.

Hypophosphatemia: A below normal level of phosphate in the blood. Normal blood levels are 2.6-4.6 mg/dL.

Hypotension: A below normal blood pressure. This condition can be created during a dialysis treatment if the rate of fluid removal from the bloodstream exceeds the vascular refilling rate from the interstitial area.

Hypotonic: A solution that contains a lower concentration of solutes than another solution.

Hypovolemia: The depletion of water in the intravascular fluid compartment.

Hypovolemic shock: Results from an insufficient amount of circulating blood to maintain adequate cardiac output, blood pressure and oxygenation of the tissues.

IDPN: Intra Dialytic Parenteral Nutrition.

Immunosuppressive drugs: Used to reduce the severity of immune reactions to such substances as protein.

In Vitro: A test done not in living organisms, but in an artificial environment referring to a process or reaction occurring in a test tube or culture media.

In Vivo: A test done in a patient or in a living body.

Infarction: A limited area of tissue destruction resulting from the loss of blood to that area.

Infiltration: Abnormal leakage of a substance into body tissues. In hemodialysis, it can be caused when a needle pierces a vessel wall allowing blood to move into the surrounding tissue.

Infuse: To introduce a fluid into something.

INR: International Normalized Ratio.

Interdialytic: Between dialysis treatments.

Interstitial fluid: The body fluid which surrounds the cells.

Intima: The inner lining of blood vessels.

Intracellular fluid: The fluid which is found inside the cells of the body.

Intradermal: Within the skin.

Intramuscular: Within a muscle.

Intravascular fluid: Fluid found within the blood vessels. Approximately 6% of an individual's body weight is comprised of this fluid.

Intravenous: Within a vein.

Ion: An atom or group of atoms carrying a charge of electricity by virtue of having gained or lost one or more valence electrons. Those charged with negative electricity , which travel toward a positive pole (Anode), are called anions; those charged with positive electricity, which travel towards a negative pole (Cathode), are called cations.

Ion exchange: A process by which certain ionized chemicals present in water are replaced with other ionized chemicals temporarily attached to resin particles. The exchange process is made only for ions having the similar charge.

IPD: Intermittent Peritoneal Dialysis, usually administered three times a week.

Ischemic: Local and temporary loss of circulating blood due to obstruction of circulation to a part.

Isotonic: Having equal tension, denoting solutions possessing the same osmotic pressure; more specifically, limited to a solution in which cells neither swell nor shrink.

JAS: Juxta Anastomotic Stenosis

JCAHO: Joint Commission Accreditation of Health care Organizations.

KDOQI: Kidney Disease Outcome Quality Initiative.

Kinetic: Relating to motion or movement.

Kinetic modeling: See Urea Kinetic Modeling.

K_oA: Mass transfer urea coefficient. The efficiency of a dialyzer in removing urea can be described by a constant referred to as K_oA. The product of K_o and membrane surface area (A) gives the mass transport coefficient (K_oA) of a given dialyzer expressed as mL/minute, the higher this value the more permeable the membrane.

Kt/V: Formula used to determine prescription of dialysis (K = dialyzer clearance of urea, t = dialysis time in minutes, V = volume of distribution of urea in body fluid).

K_{UF}: The ultrafiltration coefficient, which can range from 0.5 to 80 mL/hr/mmHg, depending on the membrane.

Labile: Unstable or easily changeable.

Lacrimation: The secretion and discharge of tears.

LAL: *Limulus amebocyte lysate* - a reagent used in vitro test for gram-negative bacterial endotoxin. This reagent is made from the blood of the horseshoe crabs which forms a gel or clot in the presence of bacterial endotoxin. There is an endotoxin limit in the pharmaceutical industry for U.S.P. Water for injection (WFI) of 0.25 EU/ml.

Langelier Saturation Index (LSI): A calculation that allows the prediction of whether water will precipitate or dissolve calcium carbonate in specific conditions, which forms scaling on the walls of pipes and R.O. Membrane.

Lateral: To one side or the other.

Lesion: An injury or wound or local area of degeneration.

Leukocyte: White blood cell.

Lidocaine: A medication used to numb the patient's access prior to the insertion of needles.

Lipid: A group of substances including fats, fatty acids and steroids. They are nonsoluble in water and used by the body as a fuel supply. Many can cross a dialyzer membrance and deposit themselves on components in the downstream lines of dialysate equipment. Removal requires the use of bleach or other high pH solvents.

Lipopolysaccharide (LPS): Group of structural molecules unique to the outer membrane of gram-negative bacteria.

Loop of Henle: The section of the nephron that forms a loop between the descending proximal tubule and the ascending distal tubule.

Low level disinfection: Physical or germicidal procedure that results in lowering a microbial population.

Lumen: The open space within a tube or container.

Lymphocytes: Clear leukocytes that are produced by the lymph glands. They are the most complex of the blood's white cells. They provide specific immune response when they encounter antigens in the body. They are activated when they recognize foreign matter as not being part of the human body.

Lyse: To destroy or break up cells.

Macrocyte: An erythrocyte that is larger than normal. Size range is between 0 and 12 microns in diameter.

Medial: Toward the middle or midline.

Medulla: Term used to describe the innermost part of an organ. In the kidney, it is the location of the collecting tubules and the loops of Henle.

Megohm: One million ohms. Usually used as a measure of water purity (deionized water quality), on resistivity in megohm/cm.

Metabolic Acidosis: Decreased pH and bicarbonate concentration in the body caused by the accumulation of acids.

Metabolism: Tissue change; the sum of the chemical changes occurring in tissue, consisting of anabolism, or those reactions that convert small molecules into large (e.g. amino acids to proteins) and catabolism, or those reactions that convert large molecules into small (e.g. glycogen to pyruvic acid).

Metastatic: Disease or disorder that is transferred from one organ or tissue to another area not directly related to the primary site.

Methemoglobinemia: A condition of having a higher than normal amount of methemoglobin in the blood with the result that the blood's ability to carry oxygen is decreased and the blood takes on a cyan color.

Microbial: Refers to microscopic organisms, bacteria, fungi etc.

Micro globulin: Beta-2 microglobulin is a protein (MW-11,800 d) produced by normal turnover of nucleated cells in the body. It is catabolized by normal kidney tubules. In ESRD, beta-2 micro globulin accumulates, leading to deposits of an abnormal protein- amyloid- in bone, joints, tendons and elsewhere.

Microcyte: An erythrocyte that is smaller than normal. Size is less than 5 microns in diameter.

Micromho (Microsiemens): One millionth of a Mho. Unit of measurement for conductivity (equivalent to inverse of 1 million ohms of resistance). Water quality is measured by its conductivity in micromhos per centimeter; the higher the quality the lower the conductivity reading.

Micron: (μ) A metric unit of measurement equivalent to 10^{-6} meters.(One-millionth of one meter).

Mm of Mercury (mmHg): A unit of pressure used primarily in the medical field for measuring blood pressures.

1 psi = 51.7 mmHg
1 mmHg = 133.36 Pascals
3 mmHg = 400 Pascals = 0.4 kPa

Mole (mol): One molecular weight of any given substance expressed in grams.

Molal: Solution containing 1 mole of solute in 1 kg of solvent. Molal solutions provide a definite ratio of solute to solvent molecules.

Molecular Weight: The molecular weight of a substance is, to a good approximation, the ratio between the weight of one molecule of it and the weight of a hydrogen atom. The molecular weight of hydrogen is 1.

Monovalent ion: A cation or anion having a single electrical charge.

Morbidity: The ratio of sick people to all people in a population.

Mortality: The rate of death in a population.

MRSA: Methicillin-Resistant Staphylococcus Aureus.

Mycosis: Any disease caused by a fungus.

Myocardial infarction: The interruption of blood flow to the heart tissue resulting in death of the tissue due to lack of oxygen. The patient will experience extreme chest pain which may extend down the left arm.

NANT: National Association of Nephrology Technologists.

Necrosis: Death of tissue.

Nephrectomy: The surgical removal of a kidney.

Nephron: The functional unit of the kidney.

Nephropathy: Abnormal functioning of the kidney.

Nephrotic Syndrome: Defined by a urinary protein level exceeding 3.5g per 1.73 m^2 of body surface area per day.

Net-Flux: The amount of solute leaving the blood and entering the dialysate per unit time.

Neuropathy: Inflammation and degeneration of the peripheral nerves.

Normocyte: An erythrocyte that is normal in size shape and color

Nosocomial: Relating to a hospital. Denoting a new disorder associated with being treated in a hospital.

NPCR: Normalized Protein Catabolic Rate.

NRCC: National Renal Credentialing Center.

Occlude: Close off.

Ohm: The fundamental unit of measure for electrical resistance. One thousand ohms is a kilo-ohm. One million ohms is called a megohm. A resistor with one volt applied across it will have a resistance of one ohm when a current of one ampere passes through it. The ohm is named after the German physicist Georg Simon Ohm (1787-1854).

Ohm's Law: A principal in physics which explains the relationship between voltage, electrical current and resistance. The relationship states that voltage is equal to the product of current and resistance.

Voltage = Current x Resistance

Oliguria: A daily output of less than 400 milliliters of urine. Below this value, toxic wastes begin to collect in the body.

Oncotic pressure: The pressure exerted in the capillary wall from the plasma proteins as water and solutes are removed.

Orthostatic hypotension: A drop in systolic blood pressure >15mmHg when a patient is placed in the upright position. Indicates intravascular depletion.

OSHA: Occupational Safety and Health Administration.

Osmolality: Osmotic concentration, defined as the number of osmoles. It is measured in osmol/kg water.

Osmolarity: The concentration of particles in one liter of solution.

Osmometer: This instrument is used to read the osmolality of a solution in milliosmoles (mOsm). The instrument generally measures the actual freeze point of the solution and translates this information into milliosmoles. A lowering of the freeze point of a solvent (water) by 1.86°C equals one osmole. Dialysate solutions tested using this method will yield results of 275-285 mOsm.

Osmosis: Diffusion of a solvent through a semipermeable membrane into a more concentrated solution from a less concentrated solution, until equilibrium is achieved.

Osmotic gradient: Difference in the concentration of solutes on each side of a semipermeable membrane.

Osmotic pressure: The force (pressure) resulting when two liquids having different solute concentrations are separated by a semi-permeable membrane. The osmotic pressure is the pressure exerted on a solvent by the difference between the concentrations (osmolalities) on either side of a membrane. It is a measurement of the potential energy difference between solutions on either side of a semipermeable membrane. It is created in the dialysis fluid by adding glucose which causes fluid to move out of the blood and into the dialysate.

Osteoblast: A cell that lays down new bone structure.

Osteoclast: A cell that resorbs and removes bone structure.

Osteodystrophy: Defective bone formation.

Osteomalacia: A disease characterized by a gradual softening and bending of the bones with varying severity of pain; the softening occurs because the bones contain osteoid tissue which has failed to calcify, due to lack of vitamin D or renal tubular dysfunction or by overexposure to water containing fluoride during dialysis.

Osteoporosis: Demineralization of bone increasing the possibility of bone fracture.

Oxidants: Substances that react with oxygen to destroy the walls of cells. Chlorine and Chloramines are used as oxidants to destroy micoorganisms in the water.

Ozone (O_3): An extremely active oxidizing agent which consists of three oxygen atoms. It is formed by the action of a high voltage electrical field on oxygen or air. A strong oxidizing agent, ozone can be used to disinfect storage tanks and piping loops.

Palpitation: Forcible pulsation of the heart, usually with an increase in frequency or force.

PAN: Polyacrylonitrile.

Parathyroid glands: Glands located in the neck next to the thyroid that assists in the regulation of calcium and phosphorus metabolism by secreting a hormone.

Parathyroid hormone: Also known as parathormone. This hormone is produced when the calcium level in the extracellular fluid decreases. Its action is to pull calcium from the bones. It is the main regulator of bone metabolism.

Paricalcitol (Zemplar): A synthetic analog of Vitamin D for treatment of secondary hyperparathyroidism. Paricalcitol is given intravenously or orally to ESRD patients to decrease PTH levels with minimal effect on calcium and phosphorous; however, the calcium phosphorous product should be monitored for elevations. Paricalcitol should never be used in patients with vitamin D toxicity or hypercalcemia.

Parts Per Million (PPM): A measure of concentration. One part per million is equivalent to1mg/ liter of liquid.

Patency: The state of being wide open.

Pathogenic: Causing a disease or abnormal process.

PCR: see Protein Catabolic Rate.

Percutaneous: Through the skin.

Pericardial tamponade: Compression of the heart due to increased volume of fluid in the pericardium.

Pericarditis: Inflammation of the pericardium, the sac that surrounds the heart.

Peritoneal Dialysis: The process by which sterile dialysis solution is introduced into the peritoneal cavity. The peritoneal membrane in the abdomen functions in the same way as the semipermeable membrane in the hemodialyzer.

Permeable: Allowing substances to pass through.

Petechia: A small spot or freckle formed by blood leaking into skin, usually occurs in groups (plural: petechiae).

pH: A value that represents the acidity or alkalinity of an aqueous solution. Hydrogen ion concentration of a solution is expressed as the pH value. A pH of 1 is very acidic, a pH of 7 is neutral, and a pH of 14 is very basic.

Phlebitis: Inflammation of a vein caused by trauma or introduction of an inflammatory agent such as iodine during venipuncture of a fistula.

Phlebotomy: Release of blood from a vein.

PHT: Pressure Holding Test, used to check the integrity of the reuse dialyzer.

Plasma: The fluid portion of the circulating blood without cellular elements; distinguished from the serum obtained after coagulation.

Platelet: A blood cell without a nucleus and having a number of functions related to hemostasis. It is capable of adhesion and aggregation in response to certain stimuli, and the microthrombin so formed are an important initial response to endothelial damage.

PMMA: Polymethylmethacrylate.

Pneumothorax: An accumulation of air or gas in the lungs (pleural cavity).

Polyuria: The excessive production of urine. Can be caused by increases in sugar or proteins.

Polyvalent ion: A cation or anion having a multiple electrical charge.

Pore: A very small opening or hole.

Posterior: Behind or toward the back of something.

Potable water: Water suitable for human consumption.

Potassium: This element is the main cation in intracellular fluid. It plays an important role in muscular activity in particular the heart muscle. Normal blood plasma levels are 3.5-5.0 mEq/L.

Potting compound: The substance used in hollow fibers to ensure an effective seal between blood and dialysate compartments at the end of the fibers.

Product water: The purified water stream from purification equipment, such as reverse osmosis units and ultrafilters.

Prophylaxis: The prevention of disease.

Protamine: Any of a class of low molecular weight proteins that have the characteristic of neutralizing heparin.

Protein: An essential constituent of all living cells that is formed from complex combinations of amino acids.

Protein Catabolic Rate (PCR): A patient's protein metabolism expressed in grams of protein per kilogram.

Proteinuria: A condition in which the urine contains large amounts of protein

Prothrombin time test (PTT): A simple routine hematological test of the extrinsic coagulation system which reflects changes in factors II, V, VII and X.

Proximal: Toward a center of reference. The opposite of distal.

Pruritis: Intense itching.

Pseudoaneurysm (false aneurysm): Sac or outpouching in the wall of a blood vessel.

Pseudomonas: Genus of gram negative bacteria found in soil water and decomposing material.

PSI: Pounds per square inch (Pressure).

Pulsatile: Rhythmic throbbing; A rhythmic forward thrust.

Pyrogen: A fever-causing substance. Bacterial lipopolysaccharide is one of the most potent pyrogens. If introduced into the blood stream, as little as 5 endotoxin units (1 nanogram) per kilogram of body weight causes fever in rabbits and humans. The symptoms of a pyrogen reaction range from a slight fever to high temperatures with shaking chills and severe loss of blood pressure.

Pyrolysis: A breakdown process that occurs when organic matter is subjected to elevated temperatures.

QAPI: Quality assessment and performance improvement

Qb: Indication of blood flow rate.

Qd: Indication of dialyzing fluid flow rate.

Qualitative: Identifying a substance as to kind/type.

Quantitative: Identifying a substance by amount present.

Radial: Located on the side of the forearm near the radius, the forearm bone that ends at the wrist near the base of the thumb.

RDA: Recommended Dietary Allowance.

Recovery (percent recovery): A measurement applied to reverse osmosis and ultrafiltration equipment that characterizes the ratio of product water to feed water flow rates. The measurement is descriptive of reverse osmosis or ultrafiltration equipment as a system and not of individual membrane elements. Expressed as a percentage, recovery is defined as:

% Recovery = (1 – Product concentration/Feed concentration) x 100

Rejection (percent rejection): A measure of the ability of a reverse osmosis membrane to remove salts. Expressed as a percentage, rejection is defined as:

% Rejection = (1 – Product concentration/Feed concentration) x 100

Renal: Pertaining to the kidneys.

Renin: A hormone produced in the kidney with important effects on sodium, potassium, and blood pressure regulation.

Resin: Substance capable of chemically or physically binding another substance & rendering it inactive.

Reticulocyte: Immature red blood cell.

Retrograde: In a backward manner, or opposite to the usual direction.

Reverse Filtration: Back filtration when the transmembrane pressure instead of facilitating ultrafiltration does the opposite.

Reverse Osmosis: A form of water purification treatment which utilizes the rejection characteristics of ion exclusion membranes (which repel ions). It is a membrane separation process for removing solvent from a solution. In normal osmosis, water molecules will flow from areas of less concentration to those of greater concentrations in order to establish an ionic (osmotic pressure) equilibrium. The flow of solution can be prevented by applying an opposing hydrostatic pressure to the concentrated solution. If the applied hydrostatic pressure exceeds the osmotic pressure, flow of solution will be reversed, that is, solution will flow from the concentrated to the dilute solution. Organic particles are screened out, while water passes through the micropores of the membrane surface by ultrafiltration. These pores are approximately 20 angstroms in diameter. This phenomenon is referred to as reverse osmosis.

Rinseback: Process used to return the blood to the patient at the completion of hemodialysis treatment; the amount of fluid necessary to clear a dialyzer.

Salicylate: A salt of salicylic acid or aspirin.

Scaling: The buildup of precipitated salts, such as calcium carbonate, onto the surface of the membrane, pipes and tanks. Scaling is associated with decreased flux and reduced reverse osmosis rejection rates.

Scleroderma: Chronic hardening and thickening of the skin.

Sclerosis: An unusual hardening.

SDI-Silt Density Index: A measure of the ability of water to foul a membrane or plug a filter. This Index provides a relative value of suspended matter. This test is used to determine the concentration of colloids and particles in the water treatment system. The measured values reflect the rate at which a 0.45micron filter will plug with particulate material in the source of water.

Sedimentation: The process by which solids are separated from water by gravity and deposited on the bottom of a container or basin.

Seizures: A neurological disorder leading to involuntary muscle spasms and possible loss of consciousness.

Serum: Fluid portion of blood remaining after a clot has formed.

SGOT (AST): Serum glutamic-oxaloacetic transaminase - aspartate aminotransferase- the serum level of this enzyme is increased in myocardial infraction and in diseases involving destruction of liver cells.

SGPT (ALT): Serum glutamic-pyruvic transaminase, - alanine aminotransferase.

Sieving Coefficient (SC): Amount of solute removed from a solution by convection (solvent drag).

Sodium: Sodium is the major extracellular cation. It plays a leading role in maintaining proper water levels and distribution throughout the body. Normal blood plasma levels are 135-145 mEq/L.

Sodium Modeling: A technique used mainly to prevent symptoms (such as hypotension, cramps) by using sodium to minimize fluid shifts from the extracellular compartment to the intracellular compartment and maintain vascular volume. These fluid shifts are primarily caused by differences in osmolality between these compartments. The technique consists of providing higher sodium at the beginning of the hemodialysis treatment to minimize hypotensive episodes at the start of the treatment. The sodium is reduced towards the end of the treatment so patients do not leave with a high sodium level.

Sodium Variation System: Allows the standard dialysis treatment to be modified in a manner such that the acid concentrate is allowed to change according to a predetermined profile for sodium.

Soft Water: Any water that contains less than a grain per gallon (17.1mg/L) of hardness as calcium carbonate.

Softener: Ion-exchange device that removes hardness from water by exchanging calcium and magnesium in the water for sodium ions.

Solute: That constituent of a solution that is considered to be dissolved in the other, the solvent.

Solution: Generally, a liquid consisting of a solvent and one or more solutes.

Solvent: That constituent of a solution which is present in dominantly larger amounts. It dissolves the solute.

Solvent drag: Occurs when molecules of a dissolved substance are dragged along in a solvent that passes through a semipermeable membrane. Solvent drag is also called convection.

Sorbent: An agent that acts by its adsorption effect.

Sphygmomanometer: A device used to measure blood pressure utilizing an inflatable cuff placed around the leg or arm.

Spike: A small amount of a single chemical used to increase a constituent in the concentrate for a single patient's treatment.

Steal syndrome: Occurs when more blood flows through the vascular access than to the distal limb (for example, in the hand). This causes feelings of numbness and tingling in the affected arm or leg where the access is placed due to change in blood flow.

Stenosis: A narrowing of a blood vessel.

Stents: Small, expanding metal rings that can be placed inside a fistula, graft, or blood vessels to help keep the lumen from narrowing.

Sterilant: A chemical or method which kills all microorganisms, which includes bacteria, viruses, fungi and spores.

Sterile: Free from all living organisms and variable spores, within the limits of tests for sterility, and maintained in that state by suitable means.

Subclavian vein: The large vein which extends from the first rib to the collarbone.

Subcutaneous: Beneath the skin.

Surface Area: In hemodialysis, the amount of membrane in direct contact with blood & dialysate.

Synthetic: Manmade, not occurring naturally.

Systemic: Affecting the entire body.

Systole: The period of time when the heart is contracting to empty itself of blood. *See diastole.*

Tachycardia: Excessively rapid heartbeat. Usually applies to rates over 100 per minute.

TCD: Theoretical conductivity.

TCV (Total cell volume): The volume of an aqueous liquid to fully prime the blood compartment of a hollow fiber dialyzer. This volume is the sum of the fiber bundle volume and header volume.

TDS (total dissolved solids): Sum of all ions in a solution (both organic and inorganic) often approximated by means of electrical conductivity or resistivity measurements. Total dissolved solids measurements are commonly used to assess reverse osmosis unit performance. TDS is measured by conductivity.

tPA: Tissue Plasminogen Activator. A thrombolytic used to declot or prevent clotting in a catheter.

Tetany: A condition where nerves and muscles become hyper excitable due to low concentrations of ionized calcium in the extracellular fluid. The patient will experience twitching and muscle cramps.

Thrill: The vibration of blood flowing through a patient's fistula or graft. It can be felt by touching the patient's access.

Thrombectomy: Surgery or use of a clot dissolving medication to remove thrombus or blood clot.

Thrombolysis: The process of injecting a drug to dissolve thrombus.

Thrombosis: The formation of a clot.

Thrombus: Clot formed in a blood vessel or a blood passage.

TMP (Transmembrane pressure): The pressure exerted across the semipermeable membrane. In dialysis, the TMP is the combined positive pressure on the blood compartment and negative pressure on the dialysate compartment.

TNF (Tumor Necrosis Factor): Also called cachectin, because it is released from tumors producing a syndrome, called cachexia, in cancer patients. TNF triggers the release of cytokine interleukin-1 from macrophages and endothelial cells, which in turn triggers the release of other cytokines and prostaglandins. These mediators at first defend the body against the offending, but ultimately turn against the body. The mediators act on the blood vessels and organs to produce vasodilatation, hypotension and organ system dysfunction.

Tortuous: Full of twists or turns, winding.

Total Chlorine: The total concentration of chlorine in a water sample, including combined and free chlorine.

TPE: Therapeutic plasma exchange.

TPR: Total peripheral resistance.

Transferrin: The transport form of iron, (Normal >20%).

Trauma: Injury or wound.

Trendelenburg Position: A body position in which the head is placed at a 45-degree downward incline on a table with the legs elevated.

Turbidity: Supension of fine particles in water that causes cloudiness and will not readily settle due to small particle size.

UFR (Ultrafiltration rate): The rate at which fluid moves from the blood into dialysate, through the semipermeable membrane. ($TMP \times K_{UF}$) or total target fluid to be removed ÷ hours of dialysis.

Ulnar: Toward the ulna, the forearm bone on the medial side when the arm is held in anatomic position.

Ultrafilters: A membrane based filtration system in which the pore sizes range from 0.001-0.1 microns.

Ultrafiltration: The transfer of fluid between the blood and dialysate through the dialysis membrane due to a pressure gradient (TMP) existing between the blood and dialysate compartment.

Ultrafiltration Coefficient: A number assigned to a particular artificial kidney which represents the number of milliliters of fluid which the dialyzer will remove per hour per mmHg of transmembrane pressure. The units of measure for the index are ml/hr/mmHg. The index is known to vary widely with dialyzers of the same size and model, therefore the index is used only as an estimate of dialyzer performance. Dialyzer reuse will also affect this value.

Ultrafiltration Profiling: UF profiling enhances the patient's vascular refill rate by pulling more fluid early in the treatment, and less at the end. This technique helps to decrease hypotensive episodes and cramping in patients without altering concentrate prescription.

Urea: One of the chief nitrogenous waste products formed by metabolism or breakdown of proteins in the body.

Urea Kinetic Modeling (UKM): A mathematic tool used to prescribe and monitor dialysis therapy.

Urea Reduction Ratio (URR): A method to calculate adequacy of dialysis. The percentage of urea reduction = 100 x (1- Ci/Co)

Uremia: A state of having excessive quantities of urea and other metabolic wastes in the body.

Uremic Frost: A white powder-like substance that forms on the skin of kidney patients due to higher concentrations of salts in their perspiration.

URR: Urea Reduction Ratio.

Urticaria: An allergic skin reaction often referred to as hives.

UV Sterilization: Using ultraviolet radiation to kill or inhibit the growth of bacteria by altering their genetic material (DNA) so they cannot multiply.

Valence: A whole number representing the combining power of an element or ion.

Vascular: Having to do with the blood vessels.

Vasoconstrict: To tighten the blood vessels.

Velocity: Measured in units of distance per time such as feet per second; also, a quantitative expression of the rate of linear motion at which water passes through a pipe; rate and direction of movement.

Venospasm: Involuntary contraction or narrowing of a vein stopping or reducing normal flow of blood.

Venous: Anything related to veins.

Virology: The branch of microbiology that deals with viruses and viral diseases.

Virus: Submicroscopic, infectious living agents that are causative factors of many illnesses.

Viscosity: The resistance of fluids to flow, due to internal forces and friction between molecules.

VRE: Vancomycin-Resistant Enterococci. highly infectious resistant bacteria

White Blood Cells (WBCs): Leukocytes; produced in bone marrow; act to combat infection and destroy bacteria.

Internal Structure of the Kidney

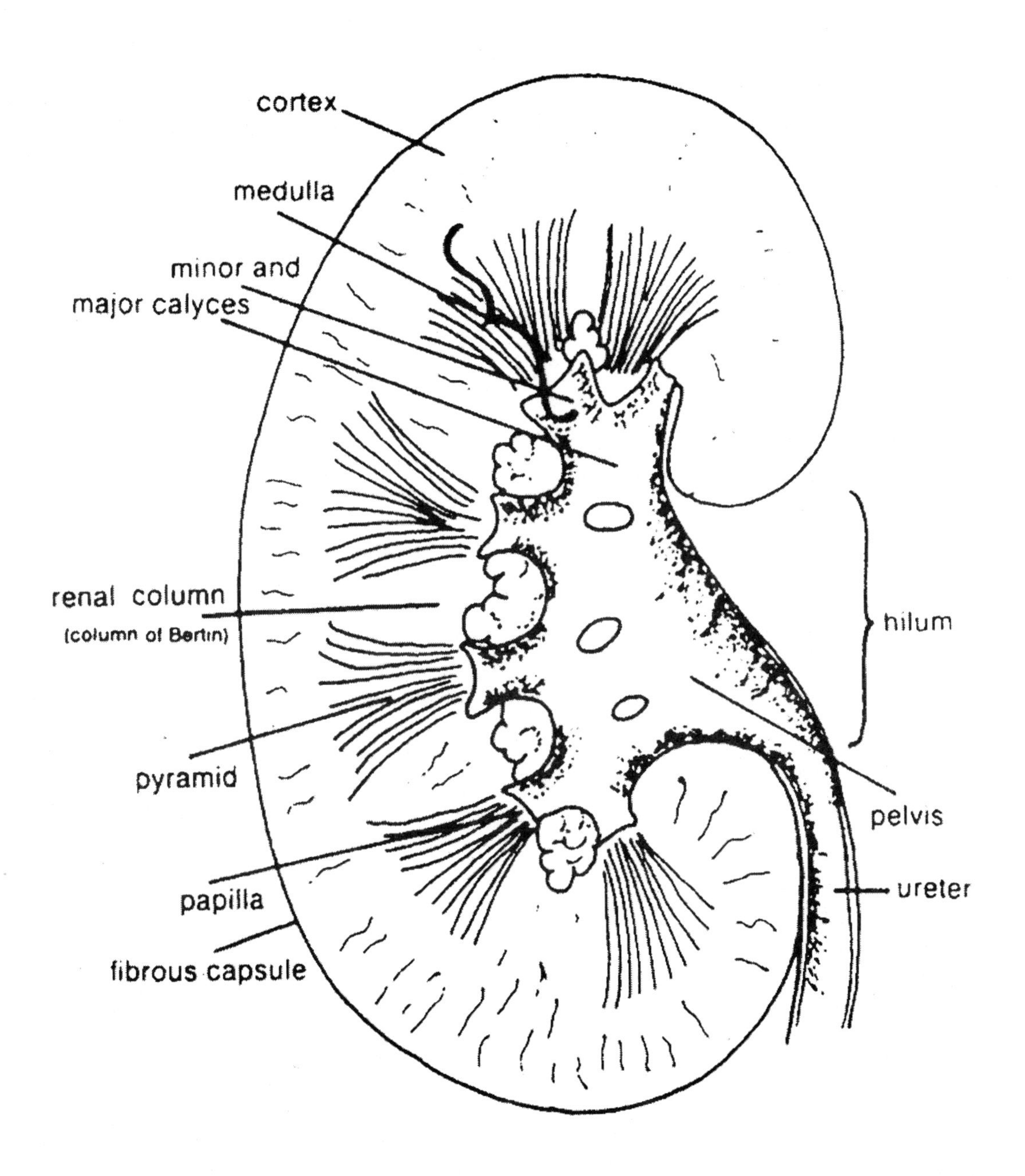

Summary of Nephron Structure

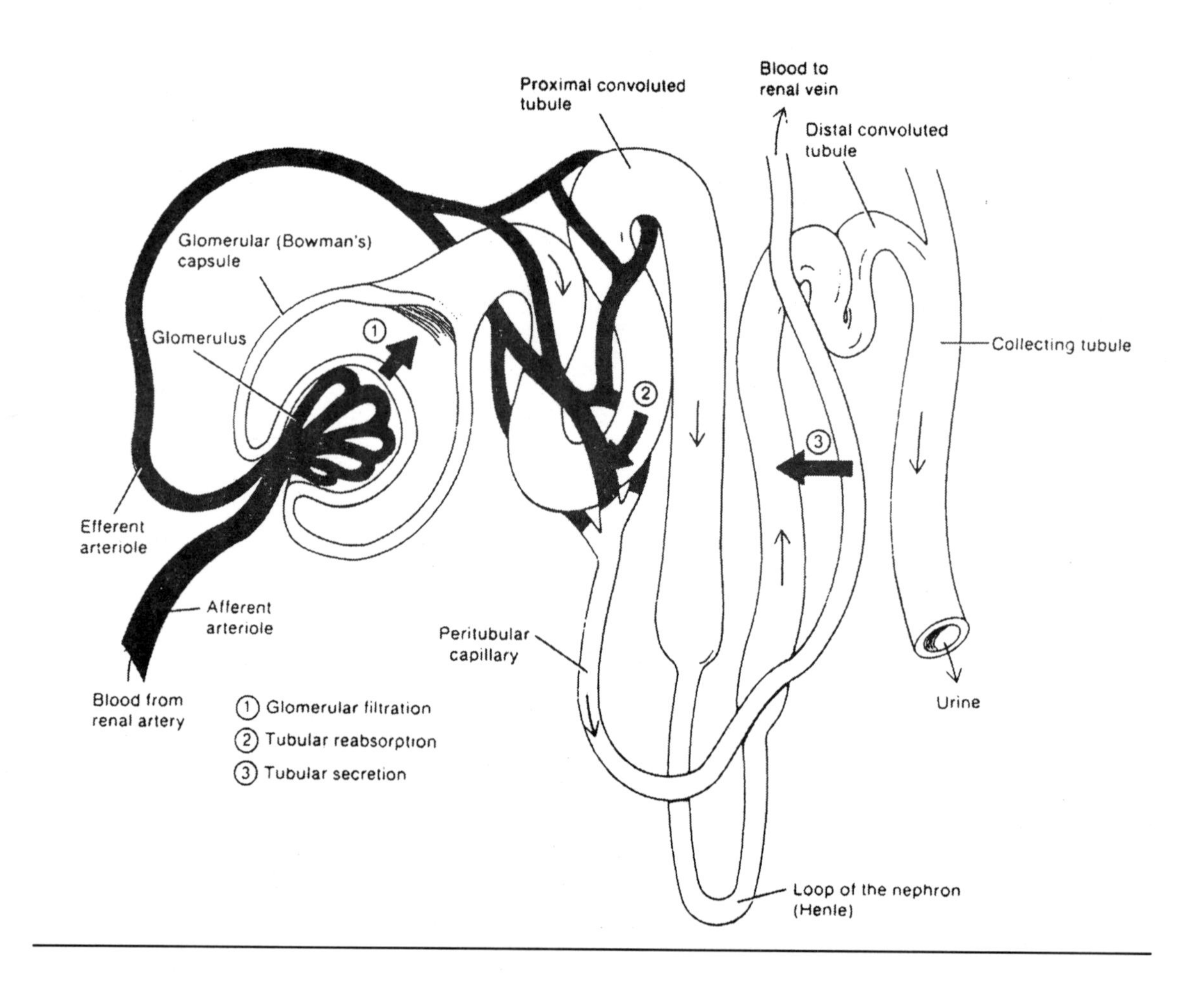

WATER TREATMENT FOR HEMODIALYSIS

Signs and Symptoms with improperly treated water used for dialysis

Contaminant	Toxic effects
Aluminum	Anemia Dialysis dementia Osteomalacia Bone diseses
Chloramine	Hemolysis Anemia Methemoglobinemia
Fluoride	Osteomalacia Osteoporosis
Calcium/Magnesium	Muscle Weakness Nausea and Vomiting
Copper	Hemolysis Liver damage
Zinc	Anemia
Nitrates	Methemoglobinemia Hypotension
Microbial/endotoxin	Pyrogenic reaction Infection
Sulfates	Nausea,Vomiting Metabolic acidosis

Components, advantages And AAMI recommendations

Component	Advantages	AAMI - Recommendation
Sediment Filters Depth Filtration Multimedia filters	Remove particulate matter	Opaque housings,(to prevent algae growth. Pressure gauges, pre and post monitor pressure drop and change filters as per manufacturers recommendation. Monitor for bacterial contamination
Water softener	- Remove Calcium and Magnesium - Protect against scaling of R.O. membrane	- Use only pellet or cubes salt designed for softeners. - Automatic regeneration and bypass - Check hardness daily before dialysis and record. - Check the regeneration timer and brine tank and record daily in the log.
Carbon Filtration	Adsorbs chlorine and chloramines	Use Granular activated carbon(GAC) - Use carbon tank in series with contact time (EBCT) of 5 minutes on each tank. - Check efferent of the first tank before every patient shift, replace tank when chloramine level >0.1mg/l - Monitor for bacteria
Reverse Osmosis	- Rejects univalent and divalent ions at the rate of 95 to 99 percent. - Remove microbiological contaminants at the rate of 99 percent	- Continuous monitor for conductivity with audible and visual alarm - Produce AAMI quality water - Monitor other RO system parameters and document the results daily. - Periodically analysis chemical(at least yearly) and bacterial contaminants,(at least monthly)
Deionization	Removes cations and anions does not remove endotoxin	- Continuous monitor resistivity more than >1m-/cm - Temperature compensated monitor - Visual and audio alarm - Carbon tank upstream - Monitor for bacteria

AAMI WATER QUALITY STANDARD

Contaminant	Recommended Maximum Level(mg/L)
Physiological substances normally included in dialysate	
Calcium	2 (0.1mEq/L)
Magnesium	4 (0.3mEq/l)
Potassium	8 (0.2mEq/l)
Sodium	70 (3.0 mEq/l)
Chemicals with maximum levels set at either the "no-transfer" level or one-tenth the maximum drinking water standard, whichever is higher.	
Arsenic	0.005
Barium	0.01
Cadmium	0.001
Chromium	0.014
Lead	0.005
Mercury	0.0002
Selenium	0.09
Silver	0.005
Chemicals with known adverse effects on hemodialysis patients	
Aluminum	0.01
Chloramines	0.10
Chlorine	0.5
Copper	0.1
Fluoride	0.20
Nitrate	2.0
Sulfate	100
Zinc	0.10

AAMI Microbiological and Endotoxin Standards for Hemodialysis Fluids.

Type of Fluid	Bacteria	Endotoxin
Product Water Action Level	<200 CFU/mL 50 CFU/mL	<2 EU/mL 1 EU/mL
Water used to rinse dialyzers	≤200CFU/mL	≤2EU/ml
Water for dialyzer germicide	≤200CFU/mL	≤2EU/ml
Dialysate Action Level	≤200CFU/mL 50 CFU/mL	<2 EU/mL 1 EU/mL
Bicarbonate concentrate**	≤200CFU/mL	
Ultra pure dialysate	<0.1 CFU/mL	0.03 EU/mL

**Dialysate components may interfere with endotoxin assays*
***Manufacturer performs endotoxin assay.*

Disinfection Agent (Chemicals) Dilution Chart

Name of Chemical	Concentration Desired	For every *liter* of water add...	For every *Gallon* of water add...
Sodium Hypochlorite (Bleach 6%)	1 PPM	0.017mL	0.064 mL
	10 PPM	0.17mL	0.64 mL
	100 PPM	1.7mL	6.4 mL
	200 PPM	3.4mL	12.7 mL
	500 PPM	8.5mL	31.8 mL
Formalin (37% Formaldehyde)	2%	54.0 mL	204.0 mL
	4%	108.0 mL	409.0 mL
Peracetic Acid (Brite Rinse)	100 PPM	2 Ml	8 mL
Hydrogen Peroxide (30% H_2O_2)	3%	90 mL	341 mL
	4%	120 mL	454 mL
	5%	150 mL	568 mL
	6%	180 mL	681 mL
Renalin	1:49	20 mL	80
Citric Acid	10%	100 mg	378.5 mg

Formula: $C_i V_i = C_f V_f$, C_i - Initial concentration; V_i - initial Volume; V_f- Final volume; C_f - Final concentration

Dialysate Additive Reference

Desired ion	Ionic Valence (mEq/mmol)	Dialysate Additive	Nomenclature Name of Compound	Molecular Weight (g/mole)	Weight of 1 mEq (mg)
Sodium	$Na^{+} = 1$	NaCl	Sodium Chloride	58.44	58.44
Potassium	$K^{+} = 1$	KCL	Potassium Chloride	74.55	74.55
Calcium	$Ca^{2+} = 2$	$CaCl_{2}.2H_{2}O$	Calcium Chloride	147.02	73.50
Magnesium	$Mg^{2+} = 2$	$MgCl_{2}.6H_{2}O$	Magnesium Chloride	203.30	101.65
Acetate	$CH_{3}COO^{-} = 1$	$CH_{3}COOH$	Acetic Acid	60.05	60.05
Bicarbonate	$HCO_{3} = 1$	$NaHCO_{3}$	Sodium Bicarbonate	84.01	84.01

Dialysate Additive Guide

Additive	Desired (mEq/L)	35X 3.43L Add Grams	35X 3.78L Add Grams	36.83X 3.43L Add Grams	36.83X 3.78L Add Grams	45X 3.43L Add Grams	45X 3.78L Add Grams
Potassium Chloride	0.5	4.5	4.9	4.7	5.2	5.8	6.3
	1.0	9.0	9.8	9.4	10.4	11.5	12.7
	2.0	18.0	19.6	18.8	20.8	23	25.4
Calcium Chloride	0.5	4.4	4.9	4.6	5.1	5.7	6.3
	1.0	8.8	9.8	9.3	10.2	11.3	12.6

Fresenius Proportioning: 1:34(35X);
-For every 35 ml of final dialysate , 1 part is acid concentrate; 1.225 parts are bicarbonate concentrate, and 32.775 parts are purified water.
-Actual dilution of bicarbonate concentrate is 1:27.57
Baxter/Gambro Proportioning: 36.83X. 1:35.83
-For every 36.83 ml of final dialysate 1parts is acid concentrate 1.83 parts are bicarbonate and 34 parts are purified water.
-Actual dilution of bicarbonate concentrate is 1:19.13
Cobe Proportioning: 45X. 1:44
-For every 45 ml of final dialysate 1 part is acid concentrate; 1.72 parts are bicarbonate and 42.28 parts are purified water.
-Actual dilution of bicarbonate concentrate is: 1:25.14

Main Categories for water contamination

ORGANIC	INORGANIC	MICROBIOLOGICAL
Contain element carbon	Minerals, salt deposits	Bacteria, viruses, yeasts, molds spores
Found in plants and animals	Soil, Electrolytes, trace elements, arsenic, barium	Gram negative bacteria
Chloramines, proteins,pesticides,herbicide		Pseudomonas. non tuberculoses mycobacteria

Conversion factors for units of concentration

A	B	To convert from A to B, multiply A by	To convert from B to A, multiply B by
mg/dL	ppm	10	0.1
mg/dL	g/L	0.01	100
mg/L	%	0.0001	10000
mg/L	PPM	1	1
g/L	%	0.1	10

Conversion factors for units of flow rate

A	B	To convert from A to B, multiply A by	To convert from B to A, multiply B by
g.p.m	L/min	3.785	0.2642
ft3/min	L/min	28.32	0.0353
ft3/min	g.p.m	7.481	0.1337

Conversion factors for units of length

A	B	To convert from A to B, multiply A by	To convert from B to A, multiply B by
cm	in	0.3937	2.54
cm	ft	0.03281	30.48
M	ft	3.281	0.3048

Conversion factors for units of mass

A	B	To convert from A to B, multiply A by	To convert from B to A, multiply B by
g	lb	0.0022	453.6
kg	lb	2.205	0.4536

Conversion factors for units of volume

A	B	To convert from A to B, multiply A by	To convert from B to A, multiply B by
gal	L	3.785	0.2642
gal	ft3	0.1337	7.481
ft3	L	28.32	0.03532

Measurement Unit Conversion Chart

1 fluid ounce = 29.57 mL
1 liter = 0.2642 gallons
1 pound = 453.6 gm
1 kg = 2.2 lb
1 ml = 0.0338 fl.oz
1 gal = 128 fl.oz =3785 mL = 3.7851 L

ABBREVIATIONS	
ac	before meals
pc	after meals
ad lib	as desired
BID	twice daily
TID	3 times per day
Q.I.D	4 times per day
N/V	nausea-vomiting
c	with
PO	by mouth
NPO	nothing by mouth
prn	as needed
Stat	immediately
EDW	Estimated dry weight
SOB	shortness of breath

NORMAL VALUES

Test	Normal Value
Sodium	134-146 mEq/L
Potassium	3.5-5.1 mEq/L
Chloride	92-109 mEq/L
Bicarbonate	24-31 mEq/L
BUN	8-25 mg/dl
Creatinine	0.5-1.5 mg/dl
Glucose	60-110 mg/dl
Calcium	8.0-10.4 mg/dl
Calcium, ionized	4.25-5.25 mg/dl
Phosphorus	2.6-4.6 mg/dl
Uric acid	2.4-7.5 mg/dl
Total protein	5.6-8.4 gm/dl
Albumin	3.4-5.4 gm/dl
Alk phosphate	25-115 u/L
SGOT, AST	0/40 u/L
SGPT, ALT	0-40 u/L
CPK	5-200 u/L
Cholesterol	<200 mg/dl
LDL Cholesterol	<130 mg/dl
HDL Cholesterol	>35-40 mg/dl
Magnesium	1.6-3.0 mg/dl
Osmolality	274-296 mOsm/kg
Iron	50-160 g/dl
Iron % saturation	20-55%
Ferritin	30-250 ng/mL
Anion gap	8-12 mEq/L
Vitamin B_{12}	200-1000pg/mL

Test	Normal Value
Hemoglobin (M)	14-18 g/dl
Hemoglobin (F)	12-16 g/dl
Hematocrit (M)	40-52%
Hematocrit (F)	37-47%

Arterial Blood Gas

Test	Normal Value
pH	7.35-7.45
$PaCo_2$	35-45 mmHg
PaO_2	80-100 mmHg
HCO_3	22-28 mEq/L
O_2 Saturation, Arterial	95-98%
O_2 Saturation, Venous	60-85%

Urine

Test	Normal Value
Albumin	20-100 mg/day
Calcium	<300 mg/day
Creatinine	0.75-1.5 gm/day
Glucose	<300 mg/day
Potassium	25-115 mEq/day
Protein	10-200 mg/day
Sodium	50-250 mEq/day
Total Volume	700-1800 mL/day
Urea nitrogen	10-20 g/day
Uric acid	50-700 mg/day

Specific Gravity

Test	Normal Value
Urine	1.002-1.030
Pure Water	1.000
HCO_3 Concentrate	1.047 (45x & 35x)
HCO_3 Concentrate	1.060 (36.83x)
Acidified Concentrate	1.150 (36.83x)

Dialyzer Pressure Dynamics

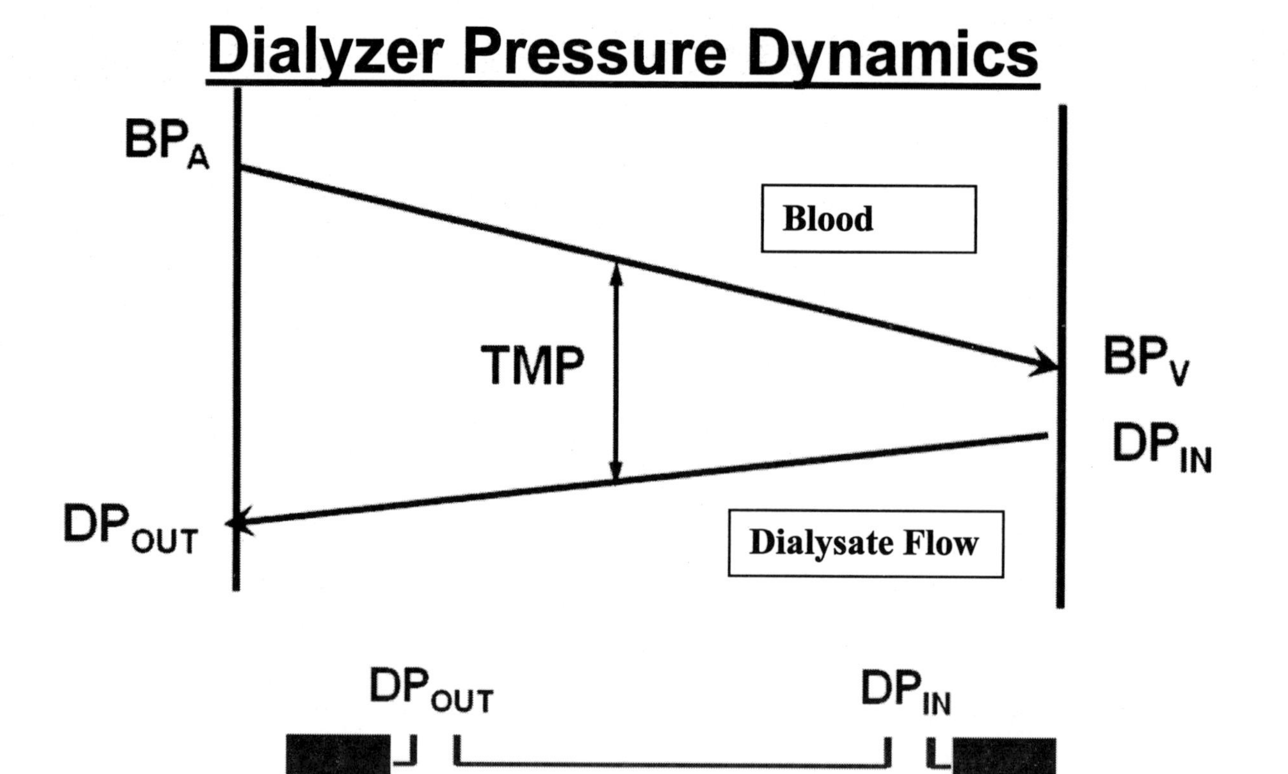

Fluids always move from areas of high pressure to areas of low pressure.

The rate of fluid movement is directly proportional to the pressure difference pressure.

The pressure difference for a dialyzer membrane is:

$$TMP = (BP_A + BP_V)/2 - (DP_{IN} + DP_{OUT})/2$$

$(BP_A + BP_V)/2$ = The average blood pressure in the dialyzer.
$(DP_{IN} + DP_{OUT})/2$ = The average dialysate pressure in the dialyzer.

FORMULAS

The symbols used for the following five formulas are:

K = *Clearance (mL/min)*
C_{bi} = *Concentration of solute in fluid entering the blood compartment (mg/dL)*
C_{bo} = *Concentration of solute in fluid leaving the blood compartment (mg/dL)*
C_{di} = *Concentration of solute in fluid entering the dialysis compartment (mg/dL)*
C_{do} = *Concentration of solute in fluid leaving the dialysis compartment (mg/dL)*
Q_{bi} = *flow rate of fluid entering the blood compartment (mL/min)*
Q_b = *flow rate of fluid entering/leaving the blood compartment with UFR = 0 (mL/min)*
Q_{di} = *flow rate of fluid entering the dialysate compartment (mg/dL)*
Q_d = *flow rate of fluid entering/leaving the dialysate compartment with UFR= 0 (mL/min)*
Q_{uf} = *ultrafiltration rate (mL/min)*
P_{bi} = *pressure at the arterial port of the hemodialyzer blood compartment (mmHg)*
P_{bo} = *pressure at the venous port of the hemodialyzer blood compartment (mmHg)*
P_{di} = *pressure at the inlet port of the hemodialyzer dialysate compartment (mmHg)*
P_{do} = *pressure at the outlet port of the hemodialyzer dialysate compartment (mmHg)*

The formulae used to calculate clearance based on the transfer of solute from the blood compartment, with and without ultrafiltration, respectively are:

FORMULA #1

$$K = \frac{C_{bi} - C_{bo}}{C_{bi}} \times Q_b$$

FORMULA #2

$$K = \frac{C_{bi} - C_{bo}}{C_{bi}} \times Q_{bi} + \frac{C_{bo}}{C_{bi}} \times Q_{uf}$$

The formulae used to calculate clearance based on the transfer of solute into the dialysate compartment, with and without ultrafiltration:

FORMULA #3

$$K = \frac{Cd_o}{C_{bi}} \times Q_d$$

FORMULA #4

$$K = \frac{C_{do}}{C_{bi}} \times Q_{di} + \frac{C_{do}}{C_{bi}} \times Q_{uf}$$

The formula used to calculate the transmembrane pressure (TMP) for a dialyzer:

FORMULA #5

$$TMP = \frac{P_{bi} + P_{bo}}{2} - \frac{P_{di} + P_{do}}{2}$$

FORMULA # 6 ***Pressure Relationships***

1 Atmosphere = 1.013 BAR = 14.7 PSI = 760 mmHg = 1.033 kg/cm^2

FORMULA # 7 ***Calculation of Urea Reduction Ratio (URR)***

$$URR = 100 \times (1 - \frac{C_t}{C_o})$$

or:

$$URR\% = \frac{preUrea - postUrea}{preUrea} \times 100$$

where:

C_t is the post dialysis BUN (mg/dL)
C_o is the pre dialysis BUN (mg/dL)

FORMULA # 8 ***Hardness Calculation***

Hardness ($CaCO_3$ mg/L) = 2.497 x Ca^{++} (mg/L) + 4.118 x Mg^{++} (mg/L)

1 grain /gal = 17.1 mg/L and 1 pound = 7000 grains.

FORMULA # 9 Empty Bed Contact Time volume (ft^3)

$$V = \frac{Q \times EBCT}{7.48} \quad or \quad EBCT = \frac{Vx7.48}{Q}$$

The required dimensions of a carbon bed can be estimated from the water flow and recommended EBCT (empty bed contact time) in minutes where Q is the water flow rate (gal/min) and 7.48 is the number of gallons in one cubic foot.

FORMULA # 10 Percent rejection for a reverse osmosis machine

$$\%rejection = 1 - \left(\frac{product water concentration}{feed water oncentration}\right) \times 100$$

or

$$\%\text{ Rejection} = \frac{PermeateFlow}{FeedFlow(permeate + rejectflow)} \times 100 =$$

Example: Your RO system is producing 8 ppm product (permeate) water and has a feed water concentration of 200 ppm:

1 - (8/200) x 100 = 96 % Rejection

FORMULA # 11 Percent Recovery for a reverse osmosis machine

$$\%\text{ Recovery} = \frac{PermeateFlow}{FeedFlow(permeate + rejectflow)} \times 100$$

Example: Your RO system is producing 4 liters per minute of permeate and 6 liters of reject water per minute.

(4.0/(6.0+4.0)) X 100 = 40% Recovery.

Note: For any RO system, the product of the feed water flowrate times its

TDS (ppm) will always equal the product of the permeate flowrate times its TDS (ppm) plus the rejection water flowrate times its TDS (ppm).

FORMULA # 12 Chloramine/Chlorine Relationship

Chloramine = Total chlorine – Free chlorine

FORMULA # 13 Urea based measurement of Recirculation

Protocol for Urea-Based Measurement of Recirculation (DOQI – Recommended)

Perform test after staring approximately 30 minutes of treatment and after turning off ultrafiltration

1-draw arterial (A) and venous (V) line samples
2-immedialtely reduce the blood flow rate to 120 ml/minute
3-turn blood pump off exactly 10 seconds after reducing blood flow rate (BFR)
4-clamp arterial line immediately above sampling port
5-draw systemic arterial sample (S) from arterial line port
6-unclamp line and resume dialysis
7-measure Bun in three samples and calculate percent recirculation
8-recirculation exceeding 10% needs to be investigated

$$R = \frac{S - A}{S - V} \times 100$$

FORMULA # 14 Mean Blood Pressure Calculation

$$\text{Mean Blood Pressure} = \frac{Systolic - Diastolic}{3} + Diastolic$$

FORMULA # 15 *Dry Powder-Weight Dilution*

Assume desired concentration of citric acid solution is 1.5%

Concentration of stock dry powder citric acid is 100%

Total amount of 1.5% solution desired is 1000mL.

-1 gram of weight is equivalent to 1 mL of fluid volume
-1.5% of the 1000 mL solution must be citric acid
-1.5% X 1000 = 15 gm of the solution must be citric acid.
-Add 15gm of powder to purified water to obtain the desired solution
-Always add the acid to the water. Start with about 800 mL and then dilute further to make the 1000 mL

-Check the calculation: $\frac{15 \text{ gm} \times 100}{1000\text{mL}} = 1.5\%$ solution

FORMULA # 16 **Stock solution at 100% concentration**

Assume desired concentration of Renalin solution is 10%
Concentration of stock solution is 100% Renalin
Total amount of 10% solution desired is 1 gallon.

- 1 gallon is equal to 128 fluid ounces
- 10% of the 128 fl.oz. of the solution must be stock Renalin.
- 10% (0.1) x 128 = 12.8 fl.oz of the solution must be Renalin and remainder is 128- 12.8 = 115.2 fl.oz of purified water (add the 12.8 fl.oz of 100% Renalin to 115.2 fl.oz. of water to obtain desired solution)
- Check the calculation: $\frac{12.8 \text{ fl.oz} \times 100}{128 \text{ fl.oz}} = 10\%$ solution

FORMULA # 17 Stock solution NOT at 100% Concentration

Assume desired concentration of hydrogen peroxide is 0.5%
Concentration of stock solution is 5% hydrogen peroxide
Total amount of 0.5% solution desired is 1000 mL.

-0.5 is what % of 5? Or 0.5/5 = 0.1 (10%)
-10% of the 1000 mL solution must be 5% stock solution to make 1000 mL of 0.5% solution
-10% (0.1) x 1000 = 100 mL of 5 % solution and the remainder is 1000 - 100 = 900 mL of purified water.
Check the calculation: $\frac{100\text{ mL }(5\%)}{1000\text{ mL}} \times 5\% = 0.5\%$ *solution*

Formula # 18 Converting Percent Concentration to Parts Per Million (PPM)

A solution consists of a solvent (the dissolving fluid) and a solute (the substance being dissolved).

100 % of a solution has 1,000,000 parts per million (ppm). A 10% solution would be 100, 000 (0.10 x 1,000,000) ppm of solute and 900,000 ppm solvent making a total of 1,000,000 ppm. The "10%" equals the amount of solute.

A 6.00% Sodium Hypochlorite solution would consist of 6% Sodium Hypochlorite and 94% water. 6% of 1,000,000 is (0.06 x 1,000,000) which equals 60,000 ppm.

A dilution calculation for Sodium Hypochlorite (bleach)

Desired Sodium Hypochlorite solution in ppm: 500 ppm

Solute to be used: 6.00% Sodium Hypochlorite solution

Final desired solution volume: 1 liter

Since 6% Sodium Hypochlorite solution is 60,000 ppm and the desired final bleach solution must be 500ppm, the 6% solution must be diluted 120 times.

60,000 ppm /500 ppm = 120

This means the final solution should be 1/120 of the 6.00% Sodium Hypochlorite solution.

A liter (1000 mL) of final solution at 500 ppm is desired, therefore:

1000 mL/120 = 8.33 mL of 6.00% Sodium Hypochlorite is needed

The final solution will be 8.33 mL of 6.00% Sodium Hypochlorite (the solute) and (1,000 – 8.33) 991.67 mL of water (the solvent) making 1,000 mL of 500 ppm Sodium Hypochlorite solution.

Checking the calculation

The concentration of the solute times its volume should equal the final concentration of the solution times it's volume

60,000 ppm x 8.33 mL = 500 ppm x 1,000 mL
60,000 x 8.33 = 500 x 1000
500,000 = 500,000

The final solution % concentration would be:

500ppm/1,000,000 ppm = 0.0005.   0.0005 x 100 = 0.05%

FORMULA # 19 *Conductivity to Resistivity conversion*

Conductivity is the reciprocal of resistivity just as conductance is the reciprocal of resistance. Conductivity is sometimes referred to as specific conductance because it is a measure of resistance per unit of length (centimeters). In the same way, resistivity is referred to as specific resistance because it is a measure of resistance times length (centimeters). The units of measure for each is:

Resistance – Ohm
Resistivity – Ohm – centimeter
Conductance – Siemens
Conductivity – Siemens/centimeter

In the dialysis field, water conductivity is measured in microSiemens/cm or resistivity in megohm-centimeters. Dialysate is almost always measured in milliSiemens/cm

To convert conductivity in microSiemens/cm to resistivity in megohm-cm, take the reciprocal by dividing the conductivity into 1.

Example: If the conductivity of a water sample is 25 microSiemens/cm, then it's resistivity would be:

$1.0/25 \text{ microSm/cm} = 1.0/25 \times 10^{-6} = 40{,}000 \text{ Ohm-cm}$
$40{,}000 \text{ Ohm-cm}/1{,}000{,}000 = 0.04 \text{ megohm-cm}$

Formula #20 Resistivity to Conductivity Conversion

To convert resistivity in megohm-cm to conductivity in micro Siemens/cm, take the reciprocal by dividing the resistivity by one.

Example: Given a resistivity of 3 megohm- cm, what is the equivalent conductivity?

$1.0/3$ *megohm-cm* $= 1.0/3 \times 10^{6} = 3.33 \times 10^{-7}$ *Siemens/cm*
3.33×10^{-7} *Siemens/cm/*$10^{-6} = 0.333$ *microSiemens/cm*

Formula #21 Conductivity to PPM Conversion

To convert conductivity measured in microSiemens/cm to the equivalent amount of dissolved solids in parts per million (ppm) multiply the conductivity in microSiemens/cm by 0.6.

CAUTION!! This formula is an approximation that works for low concentrations of dissolved solids in water such as municipal or treated water only. For water solutions such as dialysate or seawater, this ratio will not be accurate due to ion interactions at higher concentrations.

Example: If feed water has a conductivity of 200 microSiemens/cm, then:

200 microSiemens/cm x 0.6 = 120 ppm

Formula #22 — PPM to Conductivity Conversion

To convert parts per million (ppm) to a conductivity equivalent in microSiemens/cm, divide by 0.6. Remember this is an approximation for low concentrations of dissolved solids.

Example: If feed water has a total dissolved solids content of 180 ppm, what would be the expected conductivity in microSiemens/cm?

180 ppm divided by 0.6 = 300 microSiemens/cm

ESTIMATING A DIALYSATE CONDUCTIVITY

ADD TOGETHER THE CONCENTRATIONS OF SODIUM,POTASSIUM, MAGNESIUM AND CALCIUM AS STATED ON THE LABEL OF THE CONTAINER. THE CONCENTRATIONS WILL BE IN mEq/L.

SUBTRACT 6 FROM YOUR ANSWER

DIVIDE BY 10

THE ANSWER IS APPROXIMATION OF THE CONDUCTIVITY FOR THAT DIALYSATE IN mS/cm

EXAMPLE :

SUPPOSE A CONCENTRATE CONTAINER LABEL READ AS FOLLOWS:

SODIUM =137 mEq/L
CALCIUM =3.5mEq/L
POTASSIUM =2.0mEq/L
MAGNESIUM=1.0mEq/L

1. 137 + 3.5 + 2.0 + 1.0 = 143.5
2. 143.5 – 6 = 137.5
3. 137.5 ÷ 10 = 13.75
4. 13.75 mS/cm

KoA - Mass Transfer Coefficient

The number assigned to a dialyzer to indicate the permeability of the membrane for a particular molecule or particle. The most common KoA's are associated with the urea molecule. The larger the KoA value, the higher the clearance rate for the dialyzer. The units of measure for KoA are mL/min.

The KoA for a dialyzer can be determined by measuring the clearance at a fixed dialysate and blood flowrate

The formula for deriving KoA is:

$$KoA = \left[Q_B \Big/ 1 - \frac{Q_B}{Q_D} \right] \ln \left[\frac{1 - \frac{C_X}{Q_D}}{1 - \frac{C_X}{Q_B}} \right]$$

Where: C_X = Clearance of solute, X
Q_B = Blood flowrate
Q_D = Dialysate flowrate

ln = Natural logarithm

Given a KoA, a clearance can be found once Q_B and Q_D are known by the formula:

$$C_X = \frac{Q_B \left(e^{KoA\left(\frac{1}{Q_B} - \frac{1}{Q_D}\right)} - 1 \right)}{e^{KoA\left(\frac{1}{Q_B} - \frac{1}{Q_D}\right)} - \frac{Q_B}{Q_D}}$$

e = 2.718281828....

or simplified:

$$C_X = Q_B \frac{e^B - 1}{e^B - \left(\frac{Q_B}{Q_D}\right)}$$

where $B = KoA\left(\frac{1}{Q_B} - \frac{1}{Q_D}\right)$

KoA - Calculation Example

Find the KoA for the dialyzer model IP4U:

C_{BUN} = 308 ml/min **Q_B = 400 ml/min** **Q_D = 600 ml/min**

$$KoA = \left[Q_B \Big/ 1 - \frac{Q_B}{Q_D} \right] 1n \left[\frac{1 - \frac{C_X}{Q_D}}{1 - \frac{C_X}{Q_B}} \right]$$

$$= \left[400 \Big/ 1 - \frac{400}{600} \right] 1n \left[\frac{1 - \frac{308}{600}}{1 - \frac{308}{400}} \right]$$

$$= \left[400/0.333 \right] 1n \left[\frac{0.487}{0.230} \right]$$

= (1200) 1n (2.116)

= 1200 x 0.750

= 900 ml/min

Note: For $Q_B = Q_D$ neither formula can yield an answer. C_X will be equal to zero and the KoA equation cannot be worked because you can't divide by zero.

K/DOQI Stages of Kidney Disease

STAGE	***DESCRIPTION***	***GFR LEVEL(mL/min/1.73m²***
Normal kidney function	*Health Kidneys*	*90 mL/min or more*
1	*Kidney damage with normal or high GFR*	*90 mL/min or more*
2	*Kidney damage with mild decreased GFR*	*60-89 mL/min*
3	*Moderately decreased GFR*	*30-59 mL/min*
4	*Severely decreased GFR*	*15-29 mL/min*
5	*Kidney failure*	*Less than 15 mL/min or on dialysis*

PATIENT FLUID VOLUME FORMULA

METRIC SYSTEM - MALE

$$V = 0.1074\,(H) + 0.3362\,(W) - 0.09516\,(A) + 2.447$$

METRIC SYSTEM - FEMALE

$$V = 0.1069\,(H) + 0.2466\,(W) - 2.097$$

V = Volume in liters

H = Height in centimeters

W = Weight in kilograms

A = Age in years

ENGLISH SYSTEM - MALE

$$V = 0.2728\,(H) + 0.1528\,(W) - 0.09516\,(A) + 2.447$$

ENGLISH SYSTEM - FEMALE

$$V = 0.2715\,(H) + 0.1121\,(W) - 2.097$$

V = Volume in liters

H = Height in inches

W = Weight in pounds

A = Age in years

CALCULATING UREA INDEX

$$\text{UREA INDEX} = \frac{K \times t}{V}$$

K = UREA CLEARANCE RATE OF DIALYZER (mL/min)

t = TREATMENT TIME (minutes)

V = VOLUME OF BODY WATER (mL)

DETERMINE	NEED TO KNOW
PATIENT VOLUME (V)	SEX OF PATIENT HEIGHT (Inches or centimeters) WEIGHT (pounds or kilograms)
UREA CLERANCE (K)	DIALYZER K_0A BLOOD FLOW RATE (mL/min) Dialysate Flow Rate mL/min)
TIME (t)	TREATMENT TIME (minutes)
Kt/V (no units)	K (mL/min) t (min) V (mL)

ADEQUATE DIALYSIS MEANS:

$$\mathbf{1.2 \leq Kt / V \leq 1.6}$$

MEASURES ASSESSMENT TOOL (MAT) Tag Condition/Standard Measure Values Reference Source Source options: DFR=Dialysis Facility Reports **CW**=CROWNWeb **Chart**-Patient Chart **Reco** t/Staff Interview **Abbreviations:** ***CFU***=colony forming units; ***RKF***=residual kidney function; ***CHr***=reticulocyte hemoglobin; ***ESA***=erythropoiesis stimulating agent Interim Final Version 1.1 Page 1 of 3

Tag	Condition/Standard	Measure	Values	Reference	Source
494.40 Water and dialysate quality:					
V196	Water quality	Max. chloramine (must determine)	≤0.1 mg/L daily/shift	AAMI RD52	Records
V196		Max. total chlorine (may determine)	≤0.5 mg/L daily/shift		
V178		Action / Max. bacteria – product water / dialysate	50 CFU/mL / <200 CFU/mL		
V180		Action / Max. endotoxin – product water / dialysate	1 EU/mL / <2 EU/mL (endotoxin units)		
494.50 Reuse of hemodialyzers and blood lines (only applies to facilities that reuse dialyzers &/or bloodlines)					
V336	Dialyzer effectiveness	Total cell volume (hollow fiber dialyzers)	Measure original volume Discard if after reuse <80% of original	KDOQI HD Adequacy 2006; AAMI RD47	Records Interview
494.80 Patient assessment: The interdisciplinary team (IDT), patient/designee, RN, MSW, RD, physician must provide each patient with an individualized & comprehensive assessment of needs					
V502	- Health status/comorbidities	- Medical/nursing history, physical exam findings	Refer to Plan of care & QAPI sections (below) for values	Conditions for Coverage	Chart
V503	- Dialysis prescription	- Evaluate: HD every mo; PD first mo & q 4 mo			
V504	- BP & fluid management	- Interdialytic BP & wt gain, target wt, symptoms		KDOQI Hypertension & Anti-Hypertensive Agents in CKD 2004 (BP)	
V505	- Lab profile	- Monitor labs monthly & as needed			
V506	- Immunization & meds history	- Pneumococcal, hepatitis, influenza; med allergies			
V507	- Anemia (Hgb, Hct, iron stores, ESA need)	- Volume, bleeding, infection, ESA hypo-response		KDOQI HD Adequacy 2006 (volume)	
V508	- Renal bone disease	- Calcium, phosphorus, PTH & medications			
V509	- Nutritional status	- Multiple elements listed			
V510	- Psychosocial needs	- Multiple elements listed			
V511	- Dialysis access type & maintenance	- Access efficacy, fistula candidacy			
V512	- Abilities, interests, preferences, goals, desired level of participation in care, preferred modality & setting, outcomes expectations	- Reason why patient does not participate in care, reason why patient is not a home dialysis candidate			
V513	- Suitability for transplant referral	- Reason why patient is not a transplant candidate			
V514	- Family & other support systems	- Composition, history, availability, level of support			
V515	- Current physical activity level & referral to voc & physical rehab	- Abilities &barriers to independent living; achieving educational & work goals			
494.90 Plan of care The IDT must develop & implement a written, individualized comprehensive plan of care that specifies the services necessary to address the patient's needs as identified by the comprehensive assessment & changes in the patient's condition, must include measurable & expected outcomes & estimated timetables to achieve outcomes. Outcome goals must be consistent with current professionally accepted clinical practice standards.					
V543	(1) Dose of dialysis: volume	Management of volume status	Euvolemic & BP 130/80 (adult); lower of 90% of normal for age/ht/wt or 130/80 (pediatric)	KDOQI HD Adequacy 2006	Chart

V544	(1) Dose of dialysis (HD adequacy)	Adult HD <5 hours 3x/week Adult HD 2x/week, RKF <2 mL/min HD 4-6x/week	Kt/V ≥1.2; Min. 3 hours/tx if RKF <2ml/min Inadequate treatment frequency Min. Kt/V ≥2.0/week	KDOQI HD Adequacy 2006	DFR
V544	(1) Dose of dialysis (PD adequacy)	Adult PD patient <100 mL urine output/day Pediatric PD patients, low urine urea clearance	Min. delivered Kt/V_{urea} ≥1.7/week Min. delivered Kt/V_{urea} ≥1.8/week	KDOQI PD Adequacy 2006	Chart
V545	(2) Nutritional status Monitored monthly	Albumin Body weight Other parameters in Patient assessment V509	≥4.0 g/dL bromcresol green (BCG) method % usual weight, % standard weight, BMI, estimated % body fat	KDOQI Nutrition 2000 KDOQI CKD 2003	Chart
V546	(3) Mineral metabolism & renal bone disease	Calcium Phosphorus Intact PTH q 3 months	All: >8.4 mg/dL & <10.2 mg/dL All: 3.5-5.5 mg/dL Adult: 150-300 pg/mL (16.5-33.0 pmol/L) Pediatric 200-300 pg/mL	KDOQI Bone Metabolism & Disease 2003	Chart
V547 `V548 V549	(4) Anemia Monitor Hgb/Hct monthly Monitor iron stores routinely	Adult & pediatric Hgb on ESAs Adult & pediatric Hgb on ESAs Adult & pediatric Hgb off ESAs Adult & pediatric Hgb on ESAs Adult & pediatric: transferrin saturation Adult & pediatric: serum ferritin	Hgb: <12.0 g/dL[3] Hgb: 10-12.0 g/dL[4] Hgb: >10 g/dL[4] Hgb: 10-12.0 g/dL, <13.0 g/dL[5] >20% (HD, PD), or CHr >29 pg/cell[6] HD: >200 ng/mL; PD: >100 ng/mL[6] HD/PD: <500 ng/mL or evaluate if indicated[6]	[3]=FDA "black box" warning [4]=Medicare reimbursement policy [5]=KDOQI Anemia 2007 [6]=KDOQI Anemia 2006	DFR
V550 V551	(5) Vascular access	Fistula Graft Central Venous Catheter	Preferred[1,2] Acceptable if fistula not possible[1,2] Avoid, unless bridge to fistula/graft or to PD, if transplant soon, or in small adult/peds pt[1]	[1]=KDOQI Vascular Access 2006 [2]=Fistula First	DFR Interview CW
V552	(6) Psychosocial status	Survey physical & mental functioning annually KDQOL-36 survey annually	Achieve & sustain appropriate status	Conditions for Coverage CMS CPM	Chart Interview
V553 V554	(7) Modality	Home dialysis referral Transplantation referral	Candidacy or reason for non-referral	Conditions for Coverage	Chart Interview
V555	(8) Rehabilitation status	Productive activity desired by patient Pediatric: formal education needs met Vocational & physical rehab referrals as indicated	Achieve & sustain appropriate level, unspecified	Conditions for Coverage	Chart Interview
V562	(d) Patient education & training	Dialysis experience, treatment options, self-care, QOL, infection prevention, rehabilitation	Documentation of education in record	Conditions for Coverage CMS CPM 4/1/2008	Records Interview

494.110 Quality assessment & performance improvement (QAPI): The dialysis facility must develop, implement, maintain, & evaluate an effective, data-driven QAPI program with participation by the professional members of the IDT. The program must reflect the complexity of the organization & services (including those under arrangement), & must focus on indicators related to improved health outcomes & the prevention & reduction of medical errors. The dialysis facility must maintain & demonstrate

evidence of its QAPI program including continuous monitoring for CMS review.					
V629	(i) HD adequacy (monthly) (i) PD adequacy (rolling average each patient tested ≤4 months)	HD: Adult (patient with ESRD ≥3 mo) PD: Adult	% with spKt/V ≥1.2 or URR ≥65% (conventional 3 times/week dialysis) % with weekly Kt/V_{urea} ≥1.7 (dialysis+RKF)	Conditions for Coverage CMS CPM 4/1/2008 (all)	DFR Records
V630	(ii) Nutritional status	Unspecified in Conditions for Coverage & CPMs Refer to parameters in Patient assessment V509	↑ % within target range	Conditions for Coverage	Records
V631	(iii) Mineral metabolism/renal bone disease	Calcium, phosphorus, & PTH	↑ % in target range monthly	Conditions for Coverage CMS CPM 4/1/2008	Records
V632	(iv) Anemia management Patients taking ESAs &/or Patients not taking ESAs	Mean hemoglobin (patient with ESRD ≥3 mo) Mean hematocrit Serum ferritin & transferrin saturation or CHr	↑ % with mean 10-12 g/dL ↑ % with mean 30-36% Evaluate if indicated	Conditions for Coverage CMS CPM 4/1/2008 (all)	DFR Records
V633	(v) Vascular access (VA) Evaluation of VA problems, causes, solutions	Cuffed catheters > 90 days AV fistulas for dialysis using 2 needles Thrombosis episodes Infections per use-life of accesses VA patency	↓ to <10%[1] ↑ to ≥65%[1] or ≥66%[2] ↓ to <0.25/pt/yr (graft) or 0.50/pt/yr(fistula) ↓ to <1% (fistula); <10% (graft) ↑ % with fistula >3 yrs & graft >2 yrs	[1]=KDOQI 2006 [2]=Fistula First CMS CPM 4/1/2008	DFR Records CW 2/09
V634	(vi) Medical injuries & medical errors identification	Medical injuries & medical errors reporting	↓ frequency through prevention, early identification & root cause analysis	Conditions for Coverage	Records
V635	(vii) Reuse	Evaluation of reuse program including evaluation & reporting of adverse outcomes	↓ adverse outcomes	Conditions for Coverage	DFR Records
V636	(viii) Patient satisfaction & grievances	Report & analyze grievances for trends CAHPS In-Center Hemodialysis Survey available Other surveys for pediatric & home patients	Prompt resolution of patient grievances ↑ % of patients satisfied with care	Conditions for Coverage CMS CPM 4/1/2008	Records Interview
V637	(ix) Infection control	Analyze & document incidence for baselines & trends	Minimize infections & transmission of same Promote immunizations	Conditions for Coverage	DFR Records
V637	Vaccinations	Hepatitis B, influenza, & pneumococcal vaccines Influenza vaccination by facility or other provider	Documentation of education in record ↑ % of patients vaccinated on schedule ↑ % of patients receiving flu shots 10/1-3/31	Conditions for Coverage CMS CPM 4/1/2008	Records
V627	Health outcomes: Physical & mental functioning	Survey adult/pediatric patients KDQOL-36 survey annually	Achieve & sustain appropriate status ↑ % completing survey	Conditions for Coverage CMS CPM 4/1/2008	Records
V627	Health outcomes: Patient survival	Standardized mortality ratio (1.0 is average, >1.0 is worse than average, <1.0 is better than average)	↓ mortality	Conditions for Coverage CMS CPM 4/1/08	DFR

COMPLICATION QUICK GUIDE

PROBLEM	CAUSES	SIGNS/ SYMPTOMS	TREATMENT/ INTERVENTION
Air in Blood Lines • Microfoam/ Microbubbles	• Inadequate blood flow rate, causing negative pressure, pre-pump (line collapses) • A latex injection port that does not self-seal (using > 20 gauge needle) • Under-filling of saline administration set drip chamber • Improper deaeration (air removal) of dialysate fluid • Residual air left in blood pathway while priming • Introduction of air during dialysis (Normal saline, if too cold when exposed to the warm dialysate, forms a type of condensation that causes tiny air bubbles to adhere to the dialyzer membrane and sides of blood tubing.) • Not priming the heparin infusion line prior to starting priming of dialyzer • Inadequate connections pre blood pump, i.e. access to blood line, saline, monitor line connections	• "Foaming" in blood seen in blood lines or drip chambers	• Disconnect patient from the extracorporeal circuit and recirculate until air is removed.
Air Embolism (air bubbles carried by the blood stream into a vessel small enough to be blocked by the air bubble)	• Unarmed or defective air detector • Non-occluding, defective venous blood line clamp • Failure to place venous line behind venous line clamp • Careless administration of IV fluid, medications, etc. • Empty IV bag or medication bag • Air in blood lines or loose connections pre blood pump • Air leak in the blood tubing or at any connection in the extracorporeal circuit • Separation of blood lines • Very cold dialysate which contains large amounts of dissolved air that is released when warmed	• Large volume of air in venous line • Chest pain • Shortness of breath • Coughing • Cyanosis (blue-purple color of skin, lips, or nail beds) • Visual disturbances: double vision, blindness • Confusion, restlessness, fear • Slight paralysis of one side of the body • Coma • Possible cardiac arrest	• Clamp all blood lines. • Put patient on left side with head lower than feet • Attempt to aspirate air from access if possible • Initiate CPR as needed • Notify EMS if patient condition warrants • Notify physician • Monitor vital signs closely and support blood pressure with saline per access after air is removed • Administer oxygen per MD orders • Prepare to transport patient to hospital as needed • Document the incident, interventions provided, and patient response.
Hypotension (low blood pressure)	• Excessive ultrafiltration- removal of too much fluid or too rapid removal of fluid during the treatment • Antihypertensive medications • OTC or "street" drugs • Low blood volume • Low weight gain • Dehydration (i.e. vomiting and diarrhea) • Unstable cardiovascular status	• Gradual or sudden decrease in blood pressure, possibly accompanied by dizziness, nausea, vomiting, perspiration or cold clammy skin, tachycardia, loss of consciousness • An early symptom may be patients feeling quite warm, fanning themselves • Yawning • Low blood pressure at beginning of treatment • Pallor, weakness • Feeling faint • Increase in apical pulse • Feeling anxious • Sensation to move bowels	• Trendelenburg position • Minimum UFR • Normal saline bolus as needed • Osmotic agents per standing orders • Monitor vital signs closely • Adjust target loss (UFR) as needed
Nausea	• Hypotension • Disequilibrium Syndrome • Pyrogen reaction • Influenza or intestinal virus	• Nausea • Vomiting • Headache	• Assess for cause • Assess vital signs and treat as needed • Medicate per standing orders • Ensure clear airway if vomiting

PROBLEM	CAUSES	SIGNS/ SYMPTOMS	TREATMENT/ INTERVENTION
Muscle Cramps	• Rapid shifts in patient's fluid volume • Fluid removal exceeding patient's dry weight • Shift in blood chemistries, especially sodium • Fluid or electrolyte imbalance, especially depleted sodium • Hypokalemia (low potassium)	• Painful muscle spasms (usually in the extremities, hands and feet)	• Minimum UFR • Saline bolus • Administer osmotic agent per standing order • Adjust target loss if indicated • Assessment of dry weight
Headache	• Fluid shifts • Dialysis disequilibrium- a slower transfer of urea occurs from the brain tissue to the blood, so fluid is drawn into brain cells, causing swelling • Hypertension • Change in sodium level • Anxiety/ nervous tension • Decreased levels of caffeine, OTC or "street" drugs	• Pain in the head or facial area	• Assess cause • Analgesic per standing order • Monitor response • If unrelieved, notify physician
Dialysis Disequilibrium Syndrome In the brain, there is a slower transfer of urea from the brain tissue to the blood, so fluid is drawn into the brain, causing the brain cells to swell.	• Too rapid a change in serum electrolytes, pH, or osmolarity • Occurs more often in acute renal failure or when BUN values are very high ($\geq$ 150 mg/dl) • Cumulative missed treatments	• Hypertension (BP increases as treatment progresses) • Nausea and vomiting • Headache (unrelieved by analgesic) • Restlessness • Convulsions • Decreased level of consciousness • Coma • Death	• Decrease BFR(QB) and DFR(QD) • Shorten treatment time • Administer osmotic agent per standing order • Notify physician • If symptoms persist, discontinue treatment
Hypertension (high blood pressure)	• Disequilibrium Syndrome • Fluid overload • Noncompliance with blood pressure medication • Renin response (damaged kidneys may overproduce renin, raising blood pressure) • Volume overload due to excess sodium or water • Increase in effective cardiac output during the course of dialysis • Increased peripheral vascular resistance (possible side effect of EPO) • Anxiety	• Dizziness • Headache • Edema • Nausea • Vomiting • No symptoms (must monitor) • Frequently asymptomatic • High blood pressure reading • Gradual or sudden rise in blood pressure • Headache, blurring vision • Nausea, vomiting, irritability • May have no symptoms • Rapid rise in hematocrit • High systolic blood pressure • Nervousness	• Assess cause • Administer medications per standing orders or per physician order • If due to volume overload, may require sequential ultrafiltration • If due to renin release, administer saline bolus • Assess compliance with antihypertensive medications • If due to anxiety, provide reassurance to patient at regular intervals
Seizures	• Underlying seizure disorder • Dialysis Disequilibrium • Electrolyte imbalance • Delivery of improperly prepared dialysate • Hypotension • Reaction to chemical agent (germicide, disinfectant) • Hypoglycemia	• Change in level of consciousness • Jerking movements of arms and legs	• Trendelenburg position on side • Protect head • Protect access and extracorporeal circuit • Saline bolus • Minimum UFR • Check blood sugar (if indicated) • Administer medications as ordered • If unrelieved, discontinue dialysis
Angina	• Hypotension • Anemia- low hematocrit /hemoglobin • Anxiety • Cardiovascular disease	• Chest pain • Chest pressure	• Minimum UFR • Administer oxygen • Notify physician • Administer medications per standing orders or physician order • Monitor vital signs closely • If unrelieved, discontinue dialysis

PROBLEM	CAUSES	SIGNS/ SYMPTOMS	TREATMENT/ INTERVENTION
Dysrthythmia	• Rapid change in serum electrolytes, especially potassium • Hypotension • Volume excess • Low potassium level or rapid drop in potassium in conjunction with digitalis therapy • Myocardial infarction (blockage of an artery of the heart)	• Slow or rapid irregular pulse (heart rate) • Skipped or extra beats • Patient complains of "palpitations"	• Assess patient history • Monitor vital signs • Administer oxygen as needed • Minimum UFR • Support blood pressure • Notify physician • Carry out physician orders
Cardiac Arrest	• Electrolyte imbalance, especially high potassium • Dysrthythmias, Arrhythmias • Myocardial infarction (heart attack) • Cardiac tamponade (fluid build up in the pericardial sac surrounding the heart, preventing the heart from contracting) • Large air embolism • Hemolysis • Exsanguination (loss of all blood) • Hyperthermia (excessively high body temperature) • Severe hypotension	• Absence of apical or carotid pulse • Lack of spontaneous respiration • Unresponsiveness • Fibrillation	• Activate EMS • Start CPR • Return blood • Maintain patency of access • Notify physician • Administer oxygen
Hyperkalemia	• Eating too many foods high in potassium • Frequent infections or excessive breakdown of body tissue • High serum glucose in diabetics • Bleeding, particularly GI, or post surgical • Sepsis (systemic infection) • Hemolysis • Inappropriate dialysate K+ • Recent blood transfusions	• Mild intestinal cramping • Nausea • Vomiting • Diarrhea • Muscle weakness • Chest pain • Dysrthythmias • Numbness • Cramps in large muscles • Cardiac arrest	• Notify physician • Assess vital signs • Adjust dialysate potassium per physician order • Place on cardiac monitor if available • Administer oxygen if needed • Assess patient dietary intake • Initiate dialysis ASAP unless patient condition requires hospitalization
Hemolysis (Rupture of red blood cells)	Improperly diluted dialysis bath due to failure of mixing system or human error: • Failure to connect dialysate concentrate • Obstruction of concentrate source • Malfunction of concentrate proportioning pump • Faulty concentrate • Failure of conductivity monitor significant calibration error fouling of conductivity probe failure to set correct (safe) limits in a manual system failure to test with an independent meter • Bypass mechanism failure mechanical failure retrograde leak across the bypass valve • Failure of thermostat • Thermostat not set properly • Failure of high temperature sensor • High temperature sensor not set properly • Calibration error in dialysate temperature range • Major malfunction in heater cycle • Dialysate temperature monitor failure • Failure of machine to go into bypass: Mechanical failure retrograde leak across the valve • Mechanical destruction of RBCs due to excessive pressure, such as over-occlusion of the blood pump • Inadequate blood flow resulting in excessive negative arterial pressure (cumulative effect) • Chemical infusion	• "**Cranberry juice**" or "**cherry pop**" appearance of blood in venous line • Huge, rapid influx of water into circulation with dilution of plasma- water moves across the cell membrane and dilutes intracellular constituents (sodium, calcium, magnesium, chloride and protein) • Warmth in throat • Erratic blood pressure • Chest pain or tightness • Dyspnea • Anxiety • Restlessness • Throbbing headache • Nausea, vomiting, abdominal cramping, diarrhea • Seizures • Dysrthythmias- initially decreased pulse leading to rapid, thready pulse • Hyperkalemia • Patient complains of feeling hot • Skin is hot and may be dry • Headache • Delirium • Seizures • Rapid, weak respiration • Tachycardia (rapid pulse) • Hypotension • Chest pain • Dyspnea • Cardiac arrest • Hyperkalemia	• Clamp all blood lines, DO NOT return blood • Notify physician and EMS if patient condition warrants • Aspirate 10 ml of blood from venous access and maintain access patency • Monitor vital signs continually until stable • Support blood pressure as needed • Administer oxygen • Blood specimen may be spun to provide a quick assessment for hemolysis • Draw blood samples as ordered • Investigate cause of hemolysis ♦ dialysate temperature ♦ test for chlorine/chloramine ♦ independent measurement of conductivity ♦ dialysate sample to lab for electrolytes ♦ residual sterilant testing • Remove delivery system from use if it is a suspected delivery system problem, notify biomed • Document incident, interventions, and patient response

PROBLEM	CAUSES	SIGNS/ SYMPTOMS	TREATMENT/ INTERVENTION
Crenation (Dehydration of red cells) Due to hypertonic dialysate (too many electrolytes)	• Water supply diminished or shut off • Conductivity limits not set properly • Proportioning unit of the delivery system not functioning properly • Failure of conductivity monitor: significant calibration error fouling of probe failure to set correct limits on manual system • Failure of bypass mechanism	• Very dark red blood • Hypernatremia • Water flux from intracellular to extracellular space (water movement is faster than sodium shift) • Hypotension or hypertension • Headache • Nausea	• Clamp all blood lines, DO NOT return blood • Notify physician and EMS if patient condition warrants • Aspirate 10 ml of blood from venous access and maintain access patency • Monitor vital signs continually until stable • Support blood pressure as needed • Administer oxygen • Draw blood samples as ordered • Investigate cause of crenation ♦ independent measurement of conductivity ♦ dialysate sample to lab for electrolytes • Remove delivery system from use if it is a suspected delivery system problem, notify biomed • Document incident, interventions, and patient response
Fever and/ or Chills	• Infection of access or other source • Break in aseptic technique • Dialyzer contaminated with bacteria • Introduction of pyrogens (fever-producing substances) or endotoxin (byproducts of bacterial cell wall disintegration) via dialysate or reprocessed dialyzer	• Temperature over 99° F • Redness, swelling, or drainage at site of infection • Patient feels cold • Involuntary shaking chill • Temperature increase after dialysis is initiated, or temperature increase after termination of dialysis • Patient feels cold • Involuntary, shaking chill • Hypotension • Temperature rise after initiation of the dialysis treatment (usually an hour or more into treatment)	• Notify physician • Obtain cultures of blood or drainage • Administer medications as ordered • Avoid use of obviously infected access • Stop treatment, DO NOT return blood • Notify physician • Maintain patency of access • Monitor vital signs including temperature every 5 minutes until stable • Obtain cultures of blood, dialysate, and water entering the delivery system • Obtain endotoxin samples from water entering the delivery system, effluent and affluent dialysate • Discard dialyzer or give to reuse tech if reprocessed dialyzer • Prepare for transport if needed • Reinitiate treatment with new delivery system and extracorporeal circuit if patient stabilizes • Disinfect delivery system • Document incident, interventions, and patient response

PROBLEM	CAUSES	SIGNS/ SYMPTOMS	TREATMENT/ INTERVENTION
Membrane Biocompatibility Problems	• Immune response caused by a foreign substance • Hypersensitivity to the ethylene oxide gas used to sterilize some dialyzers	• Itching (pruritis) • Back pain • Chest pain • Nausea • Vague symptoms, generalized aching • Mild hypotension • Itching throughout body • Coughing, sneezing • Hives • Flushing • Acute bronchospasm (narrowing of breathing passages) • Wheezing • Vasodilation • Hypotension • Anaphylactic signs and symptoms: anxiety decreased cardiac output diarrhea chest tightness respiratory distress	• Treat symptoms • Notify physician • Stop dialysis if symptoms are not relieved • Stop dialysis, DO NOT return blood • Supportive therapy for patient symptoms • Administer oxygen • Follow anaphylactic protocol • Notify physician • Document incident, interventions, and patient response
Renalin Reaction	• Improper rinsing of reprocessed dialyzer • Not following policies and procedures for use of a reprocessed dialyzer (especially residual tests) • Inadequately primed dialysate compartment	May vary in intensity; one or more may be present: • Low back pain • Chest pain (severe) • SOB • Extreme agitation/ restlessness • Burning or hot sensation at venous access site or distal extremities • Flushed, hot sensation • Nausea, vomiting • Generalized muscle aching/ weakness • Swelling of eyes and lips • Numbness and tingling of lips • Shock • Blindness • Hemolysis or foam and bubbles in blood as it exits dialyzer	• Clamp all blood lines. DO NOT return blood • Trendelenburg position • Monitor vital signs continuously until stable • Administer oxygen as needed • Notify physician • Aspirate 10 ml of blood from venous needle • Maintain patency of access • Using fresh saline and administration set give saline as needed to support blood pressure • Draw labs as ordered by physician • Administer medications per physician order • Test dialysate for Renalin • Cap dialysate ports of dialyzer and save for possible future testing • If Renalin residual is positive, rinse delivery system until a negative residual is achieved • Assess for hemolysis using spun blood sample • Restart treatment with dry dialyzer and new blood lines if patient stabilizes • Prepare for transport as condition warrants • Document incident, interventions, and patient response
Power Failure	• Overloading of electrical circuit • Local power outage • Machine accidentally unplugged	• Power off alarms • Equipment stops running • Brief power outage until emergency generator kicks on	• Remove venous line from venous line clamp, hand crank to prevent blood clotting in extracorporeal circuit • Manually return blood, monitoring for air continuously, at direction of nurse

PROBLEM	CAUSES	SIGNS/ SYMPTOMS	TREATMENT/ INTERVENTION
Blood Loss (accidental)	• Arterial/venous blood line disconnection due to improper luer connection or loose connections • Arterial/venous needle slips out (dislodged from position) • Dialyzer leak or rupture • Clotted blood in venous blood line • Clotted blood in arterial blood line	• Blood noted on floor, chair, and/or patient clothing or blanket • Hypotension • Air entering extracorporeal circuit • Venous and/ or arterial pressure may alarm (usually a decrease in venous pressure, arterial pressure moves towards 0) • Blood detector alarms • Foamy pink to red-tinged dialysate • Venous pressure rises • Unable to return blood via venous line • Clots noted in venous drip chamber • Arterial blood line with air in it or line "jumping" • Possible decrease in venous pressure • Unable to pump blood into dialyzer • Change in arterial pressure depends on site of clot (clot pre blood pump will cause higher pre-pump negative arterial pressure, clot post pump will cause venous pressure to decrease)	• Stop blood pump and clamp blood lines • Monitor vital signs • Support blood pressure as needed • Apply pressure to access as indicated • Notify physician • Obtain labs as ordered • Return blood if possible • Test effluent dialysate for presence of blood • If positive, stop dialysis, DO NOT return blood • Maintain patency of access • Restart dialysis with new extracorporeal circuit if patient is stable • Return blood if able • Maintain patency of access • Restart dialysis with new extracorporeal circuit or blood line as indicated by delivery system • Notify physician • Obtain labs as ordered • Document incident, interventions, and patient response
Clotted Dialyzer (many dark fibers in dialyzer)	• Inadequate anticoagulation therapy or failure to follow procedure for administration of heparin • Clotting disorder • Dehydration of blood due to fluid removal without blood flow • Low blood flow rate • Frequent extracorporeal alarms Air: • Not purged during priming procedure • Introduced into extracorporeal circuit during treatment • Rising HCT/Hb in response to Epogen therapy	• Fibrin or clot formation in blood lines or drip chambers • Poor dialyzer rinse back • Rise in post-pump arterial pressure, with a decrease in venous pressure • Increasing TMP • Blood turns very dark Air entering extracorporeal circuit by: • Priming technique with air trapped in fibers, header caps, blood pump segment, venous drip chamber • Arterial access not completely primed when arterial line is connected • Low level in drip chambers • Administration of IV medication • Changing saline bag • Not clamping heparin infusion line • Any leak in the extracorporeal circuit pre blood pump • Cracked connections at access lines, blood lines • Loose saline T connection • Loss of blood flow causing high arterial negative pressure • Over-occluded blood pump • Blood more viscous (thick)	• Return blood if able • Maintain patency of access • Restart dialysis with new extracorporeal circuit • Notify physician • Obtain lab tests as ordered
Heparin Overdose	• Error in initial heparin dose • Error in heparin infusion pump setting • Heparin pump malfunctioning	• Unusual bleeding around needle site during treatment • Prolonged bleeding from puncture sites post dialysis (not related to access problem) • Unusual bleeding (GI, nose bleeds, bruising, etc) • Extended clotting time (ACT)	• Control bleeding • Notify physician • Administer medications as ordered

Access Related Complications			
PROBLEM	**CAUSES**	**SIGNS/ SYMPTOMS**	**TREATMENT/ INTERVENTION**
Poor Blood Flow (Inadequate blood flow and/or continuous collapsing when the blood pump is on.)	• Using an incorrect needle gauge • Clotted needle • Clamp on arterial line • Incorrect needle placement • Bevel of needle against wall of the vessel • Needle not in vessel • Vessel spasms • Improper blood pump speed • Clotted access • Kinks in access or blood tubing • Failed access	• Foaming appearance in blood tubing • Appearance of air in the blood lines with lowered levels in the drip chambers • Collapse of flexible drip chambers or arterial line pillows • Arterial line "jumping" • Blood pump segment collapse • Excessive pre-pump arterial negative pressure	AV Access • Decrease BFR (QB) • Reposition access extremity • Reposition needle • Check for occluded line CVC • Decrease BFR (QB) • Reposition patient • Check for occluded line • Have patient cough • Apply slight pressure to exit site • Reverse blood lines only as necessary • Fibrolytic agents per protocol or order
Recirculation (Already dialyzed venous blood mixing with arterial blood in the patient's access)	• Needles are too close together- tips less than 2 inches apart • Reversed blood lines (arterial access attached to venous blood line, venous access attached to arterial blood line) • Inadequate blood flow from the access (arterial stenosis or narrowing) or poor positioning of the arterial needle • Poor venous blood return due to: venous stenosis low flow through access tourniquet above venous needle	• Darkening of blood due to deoxygenation ("Black Blood Syndrome") • Increase of Hct • Increase in viscosity of blood during dialysis • Saline return in arterial blood line during initiation of dialysis or saline rinse of extracorporeal circuit • Decreasing Kt/V unexplained by other causes • Inappropriately high chemistries post-dialysis	• Check direction of flow through access and connection of blood lines • New cannulation with proper needle placement • Remove tourniquet from above venous access • Assess need for recirculation studies and complete as ordered • Notify physician
Needle Infiltration/ Hematoma (Blood leaking from a vessel into the surrounding soft tissue)	• Improper venipuncture technique • Movement of the extremity after needle placement, pushing the needle through the vessel wall • Movement of the needle through the vessel wall if needle not taped securely after placement • Premature cannulation of immature vascular access	• Burning and tenderness at insertion site • Immediate swelling at insertion site • Hardness of the area • Pain • Discoloration • Discomfort for several days	• Stop blood pump and clamp blood line • Assess appropriateness of needle removal • Ensure that bleeding is stopped • Follow unit protocol for treating infiltration and continuation of treatment
Excessive Bleeding	• Technical problems with venipuncture • Excessive manipulation of needle within vessel • Penetration of vessel walls at venipuncture or with abrupt movement of the limb • Repeated punctures at the same site (may cause thinning of skin and aneurysm of vessel) • Repeated punctures into dry, thin skin • Poor skin condition • Excessive heparin doses	• Unusual and/ or excess bleeding from puncture site during treatment • Prolonged time to reach hemostasis after needle removal	• Control bleeding at the site • Notify physician if bleeding cannot be controlled • Administer medication as ordered
Aneurysms True Aneurysm (ballooning of a weakened portion of a fistula vessel) False Aneurysm (a collection of blood in the layers of graft material which has no "communication" with blood in the access)	• Repeated puncture of a vessel at the same site; "one-site-itis" • Shearing of the walls of a vessel during venipuncture • Unsealed needle puncture site with hematoma formation • Repeated cannulation at the same site • Separation of layers of graft material which occurs with poor cannulation technique	• Dilation (enlargement) of vessel • Bounding pulse in area of dilation • Area possibly more sensitive to venipuncture than remaining fistula • Unusual dilation of graft anywhere along its course	• Notify physician of changes in size of aneurysm • Notify physician of changes in size of pseudoanuerysm

PROBLEM	CAUSES	SIGNS/ SYMPTOMS	TREATMENT/ INTERVENTION
Clotting (Thrombosis)	Any of the following can cause decreased blood flow through the access. Anytime there is decreased flow clotting of the access may occur. • Hypotension Recurrent orthostatic hypotension (decrease in blood pressure upon standing) Severe chronic hypotension Volume depletion in the vascular system Significant or frequent drops in blood pressure during the treatment • Excessive or prolonged pressure on the vessel by: Holding the extremity in one position for extended periods of time Prolonged direct pressure on the vessel following needle removal Use of constrictive clothing, bandages, clamps, or tape encircling arm • Inadequate arterial blood flow • Stenosis (narrowing) • Infection • Repeated infiltration of a site • Compression from hemorrhage into the tunnel of a graft • Compression from a pseudoaneurysm CVC Related: • Blood left to stagnate in catheter between treatments • Failure to follow procedure for flushing CVC and instilling indwelling heparin	• Absence of a bruit and/or thrill • Increased percent of recirculation • Increase in venous pressure • Able to aspirate only "black blood" during venipuncture • In a fistula, poor to no definition (dilation) of venous branches when tourniquet is applied • Occasional patient complaints of pain at the arterial anastomosis prior to complete thrombosis • Change in color of blood in arterial line when normal saline or diluted blood is introduced into circulation through the venous needle- "Black Blood Syndrome" • Sudden decrease in arterial blood flow and pressure or increase in venous pressure when graft is compressed between the two needles • Gradual increase in BUN and creatinine with no change in diet or muscle mass • Unable to aspirate (withdraw) blood from one and/or both sides of the catheter	• Replace fluid if due to hypotension • Review antihypertensive therapy • Notify physician of clotted access • Attempt to improve QB • Evaluate adequacy of access (venous pressure, QB, arterial pressure, etc.) • Assess access for S/S of infection and report findings, give medications as ordered • Notify physician of clotted CVC • Administer thrombolytic agent per order or protocol
Infection	• Break in aseptic technique at initiation or termination of dialysis • Bacteria spread from another infected area of the body • Poor hygiene, poor care of access • Abscess at site of sutures (temporary CVC) • Break in sterile technique when inserting central vein catheter	• Redness • Swelling • Tenderness • Drainage • Warmth • Fever (with or without other signs if the infection is in the interior part of the graft) • Hypotension • Positive blood cultures • Fever and/ or chills as hemodialysis treatment progresses	• Notify physician • Administer medication as ordered • Culture drainage • Avoid cannulation at infected site • Apply dressing as needed • Monitor vital signs
Ischemia Steal Syndrome	• Fistula: If the anastomosis is too big, it can deprive the rest of the limb of an adequate amount of blood flow. This problem may be worse in patient with vascular disease, such as diabetics. • Grafts: Normal arterial blood supply to distal extremity is shunted through the graft, thus depriving the distal extremity needed oxygenation. Occurs most commonly in patients with poor distal circulation before creation of the AV graft.	• Affected limb colder distally (furthest from the limb) than non-affected limb • Cyanotic (blue) nail beds on affected limb • Pain in distal extremity (fingers or toes) ranging from mild to severe; usually made worse or precipitated by dialysis (is the classic symptom) • Pale distal extremity	• Keep extremity warm and covered during treatment (cannot use moist heat) • Notify physician of symptoms or if symptoms become more severe • Decrease BFR(QB) if indicated • Check distal pulses

Infectious Diseases in Hemodialysis[1]

Hepatitis B

Hepatitis B is a viral infection that can damage the liver and may lead to cirrhosis and liver failure. Hepatitis B may cause liver cancer.

Some of the ways Hepatitis B may spread include:

- direct contact with the blood or body fluids of an infected person living in the same house with someone who has lifelong Hepatitis B infections
- sharing with an infected person things like razors, toothbrushes and needles for injecting drugs
- having sex with an infected person; or having multiple sexual partners
- tattooing and body piercing

In addition, a baby can get Hepatitis B from an infected mother during childbirth.

Symptoms of Hepatitis B

People with Hepatitis B may be asymptomatic or may have:

- yellowing of the skin or eyes
- loss of appetite
- nausea or vomiting
- fever
- extreme tiredness
- stomach or joint pain

Precautions for control of Hepatitis B in Dialysis Units

The following additional steps are recommended by CDC for the control of Hepatitis:

- Test all patients at admission and staff at hire for Hepatitis B surface antigen and hepatitis B antibody
- Vaccinate susceptible staff and patients (those who have not been vaccinated and who test negative for the Hepatitis B antibody)
- Continue to screen staff and patients for Hepatitis B surface antigen and Hepatitis B antibody. Screening should include susceptible staff and patients who have not yet received Hepatitis B vaccine or are in the process of being vaccinated or those who have not responded adequately to vaccine.

Patients who are positive for Hepatitis B surface antigen should be dialyzed in separate areas on dedicated machines and they should not participate in dialyzer reuse programs.

[1] **Courtesy of National Kidney Foundation**

Hepatitis C

Hepatitis C is also a viral infection that can hurt the liver and cause cirrhosis and liver failure.

Some of the ways Hepatitis C may spread include:

- direct contact with blood of an infected person
- sharing with an infected person things like razors, toothbrushes and needles for injecting drugs
- tattooing and body piercing
- possibly, by sexual contact with an infected person or having multiple sexual partners
- possibly, by living in the same house with someone who has a lifelong Hepatitis C infection

The chance that an infected mother will pass this infection to her newborn baby is much less than for Hepatitis B.

Symptoms of Hepatitis C

Most patients do not have symptoms. As many as 80 percent may become carriers of the disease, and they can infect others. Many of these people develop liver scarring and some may also develop cancer of the liver. When symptoms occur, they may be similar to flu, and may have:

- nausea
- fatigue
- loss of appetite
- fever
- headaches
- abdominal pain

Most people do not have yellowing of the skin and eyes, but this may occur along with dark urine.

Recommendations for preventing the spread of Hepatitis C in Dialysis Units

In addition to the basic infection control, CDC recommends the following steps for the control of NAB (non-A, non-B) hepatitis and Hepatitis C:

All patients should be monitored monthly for liver enzymes, alanine aminotransferase (ALT) and aspartate aminotransferase (AST), to detect any type of NAB hepatitis, including Hepatitis C. Of particular importance are causes occurring in clusters, which might indicate a problem with infection control practices. Elevations in liver enzymes are currently more sensitive indicators of acute Hepatitis C infection than is detection of Hepatitis C antibody.

Patients who are positive for Hepatitis C antibody or who have a diagnosis of NAB hepatitis do not have to be isolated or dialyzed separately on dedicated machine. They may participate in dialyzer reuse programs.

Routine screening of patients or staff for Hepatitis C antibody is not necessary for purposes of infection control. However, dialysis centers may wish to conduct serologic surveys of their patient populations to determine the prevalence of the virus in their center, and to determine medical management for patients or staff with diagnosis of Hepatitis C.

HIV and AIDS

HIV is the human immunodeficiency virus. It causes acquired immunodeficiency syndrome (AIDS). This disease causes progressive damage to the body's immune system and increases chances of getting serious infections that people normally can resist. The most common one is caused by a parasite called Pneumocystis carinii, which causes a kind of pneumonia that is difficult to treat. People with AIDS are also more likely to develop certain cancers, such as Kaposi's sarcoma. This cancer is usually limited to the skin, but may become widespread in AIDS patients, affecting the skin, lymph nodes and abdominal organs.

Like Hepatitis B and C, the AIDS virus is present in the blood of infected people. Some of the ways AIDS may be spread include:

direct contact with the blood or other body fluids of an infected person
sexual contact with an infected person
sharing with an infected person things like needles for injecting drugs

The circs can also be passed to a newborn baby from an infected mother.

Symptoms of AIDS

There may be no symptoms early in the disease. Later, the symptoms may include the following:

feeling run-down and tired
continued fever and night sweats
weight loss
dry, hacking cough
diarrhea
shortness of breath
red or purplish bruises or sores on the body or inside the mouth or nose
swollen glands in the neck, armpits or groin
an infection that causes a thick white coating on the tongue or in the throat, and possibly a sore throat
bleeding from body openings
bruising more easily than usual

The AIDS virus is much less transmissible than Hepatits B because it is less concentrated in the blood and does not survive on environmental surfaces. In the dialysis unit, the standard infection control precautions recommended by the CDC, should provide sufficient protection against transmission of the virus.

HIV antibody testing is ***not*** recommended for patients and staff for infection control purposes in dialysis units. It should not be a prerequisite for admission to hemodialysis or peritoneal dialysis programs. Patients requiring acute, chronic or transient dialysis ***should not be*** denied admission because of their HIV status. Sometimes, an administrative decision may be made to test for HIV antibody for the purposes of medical management and counseling. However, this should be accompanied by:

informed consent of patients and staff
appropriate confirmatory testing
appropriate professional counseling
confidentiality of test results

Preventing Transmissions of Blood borne Diseases

Application of exposure control through the practice of universal precautions is one of the most effective means of infection control. A summary of the current CDC recommendations for dialysis units is as follows.

1. Gloves should be worn when:
    - touching blood and body fluids or mucous membranes
    - handling items or surfaces soiled with blood or body fluids
    - performing venipuncture or other vascular access procedures
    - all invasive procedures
2. Gloves should be changed:
    - when they are blood stained
    - after handling infectious waste containers
    - after each patient contact
    - after beginning a treatment
    - beginning touching any environmental surface such as machine dials, charts and phones
3. Gowns should be worn for:
    - procedures likely to generate splashes of blood or body fluids (e.g. initiation and termination of the dialysis treatment)
    - when performing peritoneal procedures
4. Face shields and protective eyewear should be worn for:
    - procedures likely to generate droplets of blood or body fluids
    - initiating or terminating hemodialysis or peritoneal dialysis
    - troubleshooting the vascular access
5. Wear booties to protect footwear from splashes or droplets of blood or body fluids.
6. Wash hands:
    - when entering patient area
    - before gloving and after gloves are removed
    - before leaving the patient area
    - after touching an environmental surface (e.g. clipboard or the dialysis machine without having gloved first)
7. To avoid penetration by sharps:
    - never recap needles
    - dispose of needles in puncture-resistant, color coded containers
8. In your personal hygiene, you should make sure to:
    - follow the unit's hand washing policy
    - never eat, smoke, apply cosmetics or handle contact lenses in treatment areas
9. To avoid environmental contamination:
    - maintain separate areas for clean (e.g. medication prep.) and soiled (e.g. blood samples)
    - maintain clean and dirty sinks
    - make sure to adequately clean and disinfect the treatment area between patient shifts (include machines, chairs, blood pressure cuffs, clipboard)
    - use a separate room and a dedicated dialysis machine and avoid dialyzer reuse for patients who are positive for hepatitis surface antigen

Recommendations for Hepatitis B Vaccination and Serologic Surveillance in Chronic Hemodialysis Patients and Staff

The Centers for Disease Control and Prevention (CDC) and the Immunization Practices Advisory Committee (ACIP) have published guidelines for protection against infection with hepatitis B virus (A1). This appendix is meant to collate, summarize, and update, but not replace, sections of these guidelines that deal specifically with hemodialysis patients and staff. If a patient or staff member is exposed to hepatitis B virus, the recommendation of the ACIP (A2) should be followed.

Initial Testing for Hepatitis B Virus Markers

Hemodialysis patients and staff should be tested for hepatitis B surface antigen (HbsAg) and antibody to HbsAg (anti-HBs) when they begin dialysis or employment in the center. They are classified as infected if HbsAg-positive; immune if anti-HBs positive (≥ 10 milli-international units per milliliter[mIU/mL]) on at least two consecutive occasions; or susceptible if HbsAg-negative and anti-HBs negative (< 10 mIU/mL).

For infection control purposes, testing for antibody to hepatitis B core antigen (anti-HBc) is not necessary. However, if testing is done, individuals who are HbsAg-negative and anti-HBc positive have had past hepatitis B virus infection and are immune.

Hepatitis B Vaccination

All susceptible patients and staff should receive hepatitis B vaccine (dosage schedules in Table A1), be tested for anti-HBs 1-2 months after the final dose of vaccine, and be followed up as outlined below. Vaccination of immune (anti-HBs ≥ 10 mIU/mL on two consecutive occasions) persons is not necessary, but also is not harmful.

Screening and Follow-up

Screening and Follow-up depends on the result of anti-HBs testing 1-2 months after the final dose of vaccine (Table A2). Unvaccinated immune individuals can be screened and followed up as if they were vaccine responders.

Patients, Responders. Patients who are anti-HBs positive (≥ 10 mIU/mL) after vaccination are responders. They should be tested for anti-HBs each year (Table A2). If the level of anti-HBs falls below 10 mIU/mL, they should receive a booster dose of hepatitis B vaccine and continue to be tested for anti-HBs each year.

Patients, Non-responders. Patients who are anti-HBs negative (< 10 mIU/mL) after vaccination are non-responders. They may be revaccinated with one or more doses of vaccine and retested for anti-HBs 1-2 months later. If they are then anti-HBs positive

Permission with CDC

(≥ 10 mIU/mL0, they can be reclassified and treated as responders (see above). If they continue to be non-responders (anti-HBs < 10 mIU/mL), they should be considered susceptible to HBV infection and tested for HbsAg every month and anti-HBs every 6 months (Table A2).

Staff, Responders. Staff who are anti-HBs positive (≥ 10 mIU/mL) after vaccination are responders. They do not need any further routine anti-HBs testing (Table A2). If exposed to blood from a patient known to be HbsAg-positive, such staff members should be tested fro anti-HBs; if still anti-HBs positive (≥ 10 mIU/mL), no further action is required; however, if they have become anti-HBs negative (< 10mIU/mL), they should receive a booster dose of vaccine.

Staff, Non-responders. Staff who are anti-HBs negative (< 10 mIU/mL) after vaccination are non-responders. At the center's discretion, they can be revaccinated with one or more doses of vaccine, and retested for anti-HBs 1-2 months later. If they then become anti-HBs positive (≥ 10 mIU/mL), they should be reclassified and treated as responders (see above). If they are not revaccinated, or are still anti-HBs negative (< 10 mIU/mL) after vaccination, they continue to be non-responders. Non-responders should be considered susceptible to HBV infection and tested for HbsAg and anti-HBs every 6 months (Table A2). If they are exposed to the blood of a person known to be HbsAg-positive, they should either receive 2 doses of hepatitis B immune globulin (HBIG), or receive 1 dose of HBIG and 1 dose of hepatitis B vaccine. They may receive similar treatment if exposed to the blood of a person known to be at high risk for hepatitis B.

Table A1
Hepatitis B Vaccine Dosage Schedules

Product/Group	Dose	Schedule
Recombivax HB		
Patients	40 µg (1 mL) *	3 doses at 0, 1 & 6 mths
Staff	10 µg (1 mL)	3 doses at 0, 1 & 6 mths
Engerix-B		
Patients	40 µg (2 mL) †	4 doses at 0, 1, 2 & 6 mths
Staff	20 µg (1 mL)	3 doses at 0, 1 & 6 mths or 4 doses at 0, 1, 2, and 12 mths

* Special formulation
† Two 1.0 mL doses administered at one site

Table A2
Recommendations for Serologic Surveillance for Hepatitis B Virus (HBV) among Patients and Staff of Chronic Hemodialysis Centers

Vaccination/Serologic Status and Frequency of Screening			
Group/Screening Test	**Vaccine Nonresponder or Susceptible ***	**Vaccine Responder or Natural Immunity †**	**Chronic HBV Infection ‡**
Patients			
HBsAg	Every month	None	Every year
Anti-HBs	Every 6 months	Every year	If HBsAG becomes negative
Staff			
HBsAg	Every 6 months	None	Every year
Anti-HBs	Every 6 months	None	If HBsAg becomes negative

* Anti-Hbs<10 mLU/mL
† Anti-Hbs>mLU/mL
‡ HBsAg positive for at least 6 months; or HBsAg positive, anti-HBc positive, IgM anti-HBc negative

References

A1. Moyer LA, Alter MJ, Favero MS. Review of hemodialysis-associated hepatitis B: revised recommendations for serologic screening. Semin Dial 1990; 3:201-4

A2. Centers for Disease Control. Hepatitis B virus: a comprehensive strategy for eliminating transmission in the United States through universal childhood vaccination. MMWR 1991;40(no. RR-13).

CDC-Recommendations for Preventing Transmission of Infections Among Chronic Hemodialysis Patients

Terms and Abbreviations Used in This Publication

Acute hepatitis B	Newly acquired symptomatic hepatitis B virus (HBV) infection.
Acute hepatitis C	Newly acquired symptomatic hepatitis C virus (HCV) infection.
ALT	Alanine aminotransferase, previously called SGPT.
Anti-HBc	Antibody to hepatitis B core antigen.
Anti-HBe	Antibody to hepatitis B e antigen.
Anti-HBs	Antibody to hepatitis B surface antigen.
Anti-HCV	Antibody to hepatitis C virus.
Anti-HDV	Antibody to hepatitis D virus.
AST	Aspartate aminotransferase, previously called SGOT.
AV	Arteriovenous.
Chronic (persistent) HBV infection	Persistent infection with HBV; characterized by detection of HBsAg >6 months after newly acquired infection.
Chronic (persistent) HCV infection	Persistent infection with HCV; characterized by detection of HCV RNA >6 months after newly acquired infection.
Chronic hepatitis B	Liver inflammation in patients with chronic HBV infection; characterized by abnormal levels of liver enzymes.
Chronic hepatitis C	Liver inflammation in patients with chronic HCV infection; characterized by abnormal levels of liver enzymes.
CNS	Coagulase negative staphylococci.
EIA	Enzyme immunoassay.
EPA	U.S. Environmental Protection Agency.
ESRD	End-stage renal disease.
FDA	U.S. Food and Drug Administration.
GISA	Glycopeptide-resistant *Staphylococcus aureus*.
HBcAg	Hepatitis B core antigen
HBeAg	Hepatitis B e antigen.
HBsAg	Hepatitis B surface antigen.
HBV	Hepatitis B virus.
HBV DNA	Hepatitis B virus deoxyribonucleic acid.
HCV	Hepatitis C virus.
HCV RNA	Hepatitis C virus ribonucleic acid.
HDV	Hepatitis D virus.
HIV	Human immunodeficiency virus.
Isolated anti-HBc	Anti-HBc positive, HBsAg negative, and anti-HBs negative.
MRSA	Methicillin-resistant *Staphylococcus aureus*.
NNIS	National Nosocomial Infections Surveillance system.
RIBA™	Recombinant immunoblot assay.

RT-PCR	Reverse transcriptase polymerase chain reaction.
SGOT	Serum glutamic-oxaloacetic transaminase, now called AST.
SGPT	Serum glutamic-pyruvic transaminase, now called ALT.
VISA	Vancomycin-intermediate *Staphylococcus aureus*.
VRE	Vancomycin-resistant enterococci.

The following CDC staff members prepared this report:

Miriam J. Alter, Ph.D.
Rob L. Lyerla, Ph.D.
Division of Viral and Rickettsial Diseases
National Center for Infectious Diseases

Jerome I. Tokars, M.D., M.P.H.
Elaine R. Miller, M.P.H.
Matthew J. Arduino, M.S., Dr.P.H.
Hospital Infections Program
National Center for Infectious Diseases

in consultation with

Lawrence Y.C. Agodoa, M.D.
National Institute of Diabetes and Digestive and Kidney Diseases
National Institutes of Health

Carolyn Y. Neuland, Ph.D.
Center for Devices and Radiologic Health
U.S. Food and Drug Administration

Summary

These recommendations replace previous recommendations for the prevention of bloodborne virus infections in hemodialysis centers and provide additional recommendations for the prevention of bacterial infections in this setting. The recommendations in this report provide guidelines for a comprehensive infection control program that includes a) infection control practices specifically designed for the hemodialysis setting, including routine serologic testing and immunization; b) surveillance; and c) training and education. Implementation of this program in hemodialysis centers will reduce opportunities for patient-to-patient transmission of infectious agents, directly or indirectly via contaminated devices, equipment and supplies, environmental surfaces, or hands of personnel. Based on available knowledge, these recommendations were developed by CDC after consultation with staff members from other federal agencies and specialists in the field who met in Atlanta on October 5--6, 1999. They are summarized in the Recommendations section. This report is intended to serve as a resource for health-care professionals, public health officials, and organizations involved in the care of patients receiving hemodialysis.

INTRODUCTION

The number of patients with end-stage renal disease treated by maintenance hemodialysis in the United States has increased sharply during the past 30 years. In 1999, more than 3,000 hemodialysis centers had >190,000 chronic hemodialysis patients and >60,000 staff members (*1*). Chronic hemodialysis patients are at high risk for infection because the process of hemodialysis requires vascular access for prolonged periods. In an environment where multiple patients receive dialysis concurrently, repeated opportunities exist for person-to-person transmission of infectious agents, directly or indirectly via contaminated devices, equipment and supplies, environmental surfaces, or hands of personnel. Furthermore, hemodialysis patients are immunosuppressed (*2*), which increases their susceptibility to infection, and they require frequent hospitalizations and surgery, which increases their opportunities for exposure to nosocomial infections.

Historically, surveillance for infections associated with chronic hemodialysis focused on viral hepatitis, particularly hepatitis B virus (HBV) infection. CDC began conducting national surveillance for hemodialysis-associated hepatitis in 1972 (*3,4*). Since 1976, this surveillance has been performed in collaboration with the Health Care Financing Administration (HCFA) during its annual facility survey. Other hemodialysisassociated diseases and practices not related to hepatitis have been included over the years (e.g., pyrogenic reactions, dialysis dementia, vascular access infections, reuse practices, vancomycin use), and the system is continually updated to collect data regarding hemodialysisassociated practices and diseases of current interest and importance (*5--18*).

Recommendations for the control of hepatitis B in hemodialysis centers were first published in 1977 (*19*), and by 1980, their widespread implementation was associated with a sharp reduction in incidence of HBV infection among both patients and staff members (*5*). In 1982, hepatitis B vaccination was recommended for all susceptible patients and staff members (*20*). However, outbreaks of both HBV and hepatitis C virus (HCV) infections continue to occur among chronic hemodialysis patients. Epidemiologic investigations have indicated substantial deficiencies in recommended infection control practices, as well as a failure to vaccinate hemodialysis patients against hepatitis B (*21,22*). These practices apparently are not being fully implemented because staff members a) are not aware of the practices and their importance, b) are confused regarding the differences between standard (i.e., universal) precautions recommended for all health-care settings and the additional precautions necessary in the hemodialysis setting, and c) believe that hepatitis B vaccine is ineffective for preventing HBV infection in chronic hemodialysis patients (*22*).

Bacterial infections, especially those involving vascular access, are the most frequent infectious complication of hemodialysis and a major cause of morbidity and mortality among hemodialysis patients (*1*). During the 1990s, the prevalence of antimicrobial-resistant bacteria (e.g., methicillin-resistant *Staphylococcus aureus* [MRSA] and vancomycin-resistant enterococci [VRE]) increased rapidly in health-care settings, including hemodialysis units (*18,23*). Although numerous outbreaks of bacterial infections in the hemodialysis setting have been reported (*24*), few studies exist regarding the epidemiology and prevention of endemically occurring bacterial infections in hemodialysis patients, and formal recommendations to prevent such infections have not been published previously. In 1999, CDC initiated a surveillance system for bloodstream and vascular access infections in outpatient hemodialysis centers to determine the frequency of and risk factors for these complications in order to formulate and evaluate strategies for control (*25*).

The recommendations contained in this report were developed by reviewing available data and are based on consultations with specialists in the field. These recommendations provide guidelines for infection control strategies, unique to the hemodialysis setting, that should be used to prevent patient-to-patient transmission of bloodborne viruses and pathogenic bacteria. They are summarized on pages 20--21.

These recommendations do not address sources of bacterial and chemical contaminants in dialysis systems, water treatment or distribution, specific procedures for reprocessing dialyzers, clinical practice methods to prevent bacterial infections (e.g., techniques for skin preparation and access), or comprehensive strategies for preventing infections among health-care workers (see Suggested Readings for information on these topics).

BACKGROUND

Hepatitis B Virus Infection

Epidemiology

Incidence and Prevalence. In 1974, the incidence of newly acquired (i.e., acute) HBV infection among chronic hemodialysis patients in the United States was 6.2%, and selected hemodialysis centers reported rates as high as 30% (*4*). By 1980, nationwide incidence among patients had decreased to 1% (*5*), and by 1999, to 0.06% (*18*)

(CDC, unpublished data, 2001), with only 3.5% of all centers reporting newly acquired infections. Prevalence of chronic HBV infection (i.e., hepatitis B surface antigen [HBsAg] positivity) among hemodialysis patients declined from 7.8% in 1976 to 3.8% in 1980 and to 0.9% by 1999 (*5,18*) (CDC, unpublished data, 2001). In 1999, a total of 27.7% of 3,483 centers provided dialysis to ≥1 patient with either acute or chronic HBV infection (CDC, unpublished data, 2001).

Transmission. HBV is transmitted by percutaneous (i.e., puncture through the skin) or permucosal (i.e., direct contact with mucous membranes) exposure to infectious blood or to body fluids that contain blood, and the chronically infected person is central to the epidemiology of HBV transmission. All HBsAg-positive persons are infectious, but those who are also positive for hepatitis B e antigen (HBeAg) circulate HBV at high titers in their blood ($10^{8\text{--}9}$ virions/mL) (*26,27*). With virus titers in blood this high, body fluids containing serum or blood also can contain high levels of HBV and are potentially infectious. Furthermore, HBV at titers of $10^{2\text{--}3}$ virions/mL can be present on environmental surfaces in the absence of any visible blood and still result in transmission (*28,29*).

HBV is relatively stable in the environment and remains viable for at least 7 days on environmental surfaces at room temperature (*29*). HBsAg has been detected in dialysis centers on clamps, scissors, dialysis machine control knobs, and doorknobs (*30*). Thus, blood-contaminated surfaces that are not routinely cleaned and disinfected represent a reservoir for HBV transmission. Dialysis staff members can transfer virus to patients from contaminated surfaces by their hands or gloves or through use of contaminated equipment and supplies (*30*).

Most HBV infection outbreaks among hemodialysis patients were caused by cross-contamination to patients via a) environmental surfaces, supplies (e.g., hemostats, clamps), or equipment that were not routinely disinfected after each use; b) multiple dose medication vials and intravenous solutions that were not used exclusively for one patient; c) medications for injection that were prepared in areas adjacent to areas where blood samples were handled; and d) staff members who simultaneously cared for both HBV-infected and susceptible patients (*21,31--35*). Once the factors that promote HBV transmission among hemodialysis patients were identified, recommendations for control were published in 1977 (*19*). These recommendations included a) serologic surveillance of patients (and staff members) for HBV infection, including monthly testing of all susceptible patients for HBsAg; b) isolation of HBsAg-positive patients in a separate room; c) assignment of staff members to HBsAg-positive patients and not to HBV-susceptible patients during the same shift; d) assignment of dialysis equipment to HBsAg-positive patients that is not shared by HBV-susceptible patients; e) assignment of a supply tray to each patient (regardless of serologic status); f) cleaning and disinfection of nondisposable items (e.g., clamps, scissors) before use on another patient; g) glove use whenever any patient or hemodialysis equipment is touched and glove changes between each patient (and station); and h) routine cleaning and disinfection of equipment and environmental surfaces.

The segregation of HBsAg-positive patients and their equipment from HBV-susceptible patients resulted in 70%--80% reductions in incidence of HBV infection among hemodialysis patients (*7,36--38*). National surveillance data for 1976--1989 indicated that incidence of HBV infection was substantially lower in hemodialysis units that isolated HBsAg-positive patients, compared with those that did not (*7,10*). The success of isolation practices in preventing transmission of HBV infection is linked to other infection control practices, including routine serological surveillance and routine cleaning and disinfection. Frequent serologic testing for HBsAg detects patients recently infected with HBV quickly so isolation procedures can be implemented before cross-contamination can occur. Environmental control by routine cleaning and disinfection procedures reduces the opportunity for cross-contamination, either directly from environmental surfaces or indirectly by hands of personnel.

Despite the current low incidence of HBV infection among hemodialysis patients, outbreaks continue to occur in chronic hemodialysis centers. Investigations of these outbreaks have documented that HBV transmission resulted from failure to use recommended infection control practices, including a) failure to routinely screen patients for HBsAg or routinely review results of testing to identify infected patients; b) assignment of staff members to the simultaneous care of infected and susceptible patients; and c) sharing of supplies, particularly multiple dose medication vials, among patients (*21*). In addition, few patients had received hepatitis B vaccine (*21*). National surveillance data have demonstrated that independent risk factors among chronic hemodialysis patients for acquiring HBV infection include the presence of ≥1 HBV-infected patient in the hemodialysis center who is not isolated, as well as a <50% hepatitis B vaccination rate among patients (*15*).

HBV infection among chronic hemodialysis patients also has been associated with hemodialysis provided in the acute-care setting (*21,39*). Transmission appeared to stem from chronically infected HBV patients who shared staff members, multiple dose medication vials, and other supplies and equipment with susceptible patients. These episodes were recognized when patients returned to their chronic hemodialysis units, and routine HBsAg testing was resumed. Transmission from HBV-infected chronic hemodialysis patients to patients undergoing hemodialysis for acute renal failure has not been documented, possibly because these patients are dialyzed for short durations and have limited exposure. However, such transmission could go unrecognized because acute renal failure patients are unlikely to be tested for HBV infection.

Clinical Features and Natural History

HBV causes both acute and chronic hepatitis. The incubation period ranges from 45--160 days (mean: 120 days), and the onset of acute disease is usually insidious. Infants, young children (aged <10 years), and immunosuppressed adults with newly acquired HBV infection are usually asymptomatic (*40*). When present, clinical symptoms and signs might include anorexia, malaise, nausea, vomiting, abdominal pain, and jaundice. Extrahepatic manifestations of disease (e.g., skin rashes, arthralgias, and arthritis) can also occur (*41*). The case fatality rate after acute hepatitis B is 0.5%--1%.

In adults with normal immune status, most (94%--98%) recover completely from newly acquired HBV infections, eliminating virus from the blood and producing neutralizing antibody that creates immunity from future infection (*40,42*). In immunosuppressed persons (including hemodialysis patients), infants, and young children, most newly acquired HBV infections result in chronic infection. Although the consequences of acute hepatitis B can be severe, most of the serious sequelae associated with the disease occur in persons in whom chronic infection develops. Although persons with chronic HBV infection are often asymptomatic, chronic liver disease develops in two-thirds of these persons, and approximately 15%--25% die prematurely from cirrhosis or liver cancer (*43--45*).

Subtypes of HBV exist, and infection or immunization with one subtype confers immunity to all subtypes. However, reinfection or reactivation of latent HBV infection has been reported among certain groups of immunosuppressed patients, including those who have undergone renal transplant and those infected with human immunodeficiency virus (HIV) (*46,47*). These patients were positive for antibody to hepatitis B core antigen (anti-HBc), with or without antibody to HBsAg (anti-HBs), and subsequently developed detectable levels of HBsAg. The frequency with which this occurs is unknown.

Monotherapy with alpha interferon or lamivudine is approved by the U.S. Food and Drug Administration (FDA) to treat patients with chronic hepatitis B (*48,49*). Although the dosage of lamivudine should be modified based on creatinine clearance in patients with renal impairment, no additional dose modification is necessary after routine hemodialysis. The emergence of lamivudine-resistant variants has caused concern regarding long-term use of this drug.

Screening and Diagnostic Tests

Serologic Assays. Several well-defined antigen-antibody systems are associated with HBV infection, including HBsAg and anti-HBs; hepatitis B core antigen (HBcAg) and anti-HBc; and HBeAg and antibody to HBeAg (anti-HBe). Serologic assays are commercially available for all of these except HBcAg because no free HBcAg circulates in blood. One or more of these serologic markers are present during different phases of HBV infection (Table 1) (*42*).

The presence of HBsAg is indicative of ongoing HBV infection and potential infectiousness. In newly infected persons, HBsAg is present in serum 30--60 days after exposure to HBV and persists for variable periods. Transient HBsAg positivity (lasting ≤18 days) can be detected in some patients during vaccination (*50,51*). Anti-HBc develops in all HBV infections, appearing at onset of symptoms or liver test abnormalities in acute HBV infection, rising rapidly to high levels, and persisting for life. Acute or recently acquired infection can be distinguished by presence of the immunoglobulin M (IgM) class of anti-HBc, which persists for approximately 6 months.

In persons who recover from HBV infection, HBsAg is eliminated from the blood, usually in 2--3 months, and anti-HBs develops during convalescence. The presence of anti-HBs indicates immunity from HBV infection. After recovery from natural infection, most persons will be positive for both anti-HBs and anti-HBc, whereas only anti-HBs develops in persons who are successfully vaccinated against hepatitis B. Persons who do not recover from HBV infection and become chronically infected remain positive for HBsAg (and anti-HBc), although a small proportion (0.3% per year) eventually clear HBsAg and might develop anti-HBs (*45*).

In some persons, the only HBV serologic marker detected is anti-HBc (i.e., isolated anti-HBc). Among most asymptomatic persons in the United States tested for HBV infection, an average of 2% (range: <0.1%--6%) test positive for isolated anti-HBc (*52*); among injecting-drug users, however, the rate is 24% (*53*). In general, the frequency of isolated anti-HBc is directly related to the frequency of previous HBV infection in the population and can have several explanations. This pattern can occur after HBV infection among persons who have recovered but whose anti-HBs levels have waned or among persons who failed to develop anti-HBs. Persons in the latter category include those who circulate HBsAg at levels not detectable by current commercial assays. However, HBV DNA has been detected in <10% of persons with isolated anti-HBc, and these persons are unlikely to be infectious to others except under unusual circumstances involving direct percutaneous exposure to large quantities of blood (e.g., transfusion) (*54*). In most persons with isolated anti-HBc, the result appears to be a false positive. Data from several studies have demonstrated that a primary anti-HBs response develops in most of these persons after a three-dose series of hepatitis B vaccine (*55,56*). No data exist on response to vaccination among hemodialysis patients with this serologic pattern.

A third antigen, HBeAg, can be detected in serum of persons with acute or chronic HBV infection. The presence of HBeAg correlates with viral replication and high levels of virus (i.e., high infectivity). Anti-HBe correlates with the loss of replicating virus and with lower levels of virus. However, all HBsAg-positive persons should be considered potentially infectious, regardless of their HBeAg or anti-HBe status.

Nucleic Acid Detection. HBV infection can be detected using qualitative or quantitative tests for HBV DNA. These tests are not FDA-approved and are most commonly used for patients being managed with antiviral therapy (*49,57*).

Hepatitis B Vaccine

Hepatitis B vaccine has been recommended for both hemodialysis patients and staff members since the vaccine became available in 1982 (*20*). By 1999, a total of 55% of patients and 88% of staff members had been vaccinated (*18*) (CDC, unpublished data, 2001). Two types of vaccine have been licensed and used in the United States: plasma-derived and recombinant. Plasma-derived vaccine is no longer available in the United States, but is produced in several countries and used in many immunization programs worldwide Recombinant vaccines available in the United States are Recombivax HB™ (Merck & Company, Inc., West Point, Pennsylvania) and Engerix-B® (SmithKline Beecham Biologicals, Philadelphia, Pennsylvania). Recombivax HB™ contains 10--40 µg of HBsAg protein per mL, whereas Engerix-B® contains 20 µg/mL.

.Primary vaccination comprises three intramuscular doses of vaccine, with the second and third doses given 1 and 6 months, respectively, after the first. An alternative schedule of four doses given at 0, 1, 2, and 12 months to persons with normal immune status or at 0, 1, 2, and 6 months to hemodialysis patients has been approved for Engerix-B®.

Immunogenicity. The recommended primary series of hepatitis B vaccine induces a protective anti-HBs response (defined as ≥10 milli-International Units [mIU]/mL) in 90%--95% of adults with normal immune status. The major determinant of vaccine response is age, with the proportion of persons developing a protective antibody response declining to 84% among adults aged >40 years and to 75% by age 60 years (*58,59*). Other host factors that contribute to decreased immunogenicity include smoking, obesity, and immune suppression. Compared with adults with normal immune status, the proportion of hemodialysis patients who develop a protective antibody response after vaccination (with higher dosages) is lower. For those who receive the three-dose schedule, the median is 64% (range: 34%--88%) (*60--65*), and for those who receive the four-dose schedule, the median is 86% (range: 40%--98%) (*66--72*). Limited data indicate that concurrent infection with HCV does not interfere with development of protective levels of antibody after vaccination, although lower titers of anti-HBs have been reported after vaccination of HCV-positive patients compared with HCV-negative patients (*65,73--75*).

Some studies have demonstrated that higher antibody response rates could be achieved by vaccinating patients with chronic renal failure before they become dialysis dependent, particularly patients with mild or moderate renal failure. After vaccination with four 20 µg doses of recombinant vaccine, a protective antibody response developed in 86% of predialysis adult patients with serum creatinine levels ≤4.0 mg/dl (mean: 2.0 mg/dl) compared with 37% of those with serum creatinine levels >4.0 mg/dl (mean: 9.5 mg/dl), only 12% of whom were predialysis patients (*76*). In an earlier study, a lower response to recombinant vaccine among predialysis patients was reported, possibly because patients with more severe renal failure were included (*77,78*).

Although no data exist on the response of pediatric hemodialysis patients to vaccination with standard pediatric doses, 75%--97% of those who received higher dosages (20 µg) on either the three- or four-dose schedule developed protective levels of anti-HBs (*79--81*). In the one study that evaluated vaccine response among children with chronic renal failure before they became dialysis dependent, high response rates were achieved after four-20 µg doses in both predialysis and dialysis-dependent patients, although predialysis patients had higher peak antibody titers (*82*).

Vaccine Efficacy. For persons with normal immune status, controlled clinical trials have demonstrated that protection from acute and chronic HBV infection is virtually complete among those who develop a protective antibody response after vaccination (*83,84*). Among hemodialysis patients, controlled clinical trials conducted in other countries demonstrated efficacy of 53%--78% after preexposure immunization (*85,86*). However, no efficacy was demonstrated in the one trial performed in the United States (*62*). When the latter trial was designed, the sample size was calculated based on an annual incidence rate among susceptible patients of 13.8% (i.e., the rate observed during 1976--1979, the period before the start of the trial). However, by the time the trial was conducted, the incidence rate had declined by >60%, and the sample size was inadequate for detecting a difference in infection rates between vaccinated and placebo groups. Although efficacy was not demonstrated in this study, no infections occurred among persons who developed and maintained protective levels of anti-HBs.

Furthermore, since the hepatitis B vaccine became available, no HBV infections have been reported among vaccinated hemodialysis patients who maintained protective levels of anti-HBs. This observation has been particularly striking during HBV infection outbreaks in this setting (*21*). In addition, a case-control study indicated that the risk for HBV infection was 70% lower among hemodialysis patients who had been vaccinated (*87*). Thus, most hemodialysis patients can be protected from hepatitis B by vaccination, and maintaining immunity among these patients reduces the frequency and costs of serologic screening (*88*).

Revaccination of Nonresponders. Among persons who do not respond to the primary three-dose series of hepatitis B vaccine, 25%--50% of those with normal immune status respond to one additional vaccine dose, and 50%--75% respond to three additional doses (*59,84*). A revaccination regimen that includes serologic testing after one or two additional doses of vaccine appears to be no more cost-effective than serologic testing performed after all three additional doses (*89*). For persons found to be nonresponders after six doses of vaccine, no data exist to indicate that additional doses would induce an antibody response. Few studies have been conducted of the effect of revaccination among hemodialysis patients who do not respond to the primary vaccine series. Response rates to revaccination varied from 40%--50% after two or three additional 40 µg intramuscular doses to 64% after four additional 10 µg intramuscular doses (*69,70,90--94*).

Antibody Persistence. Among adults with normal immune status who responded to a primary vaccine series with a protective antibody level, antibody remained above protective levels in 40%--87% of persons after 9--15 years (*95--98*). Only short-term data are available for hemodialysis patients. Among adults who responded to the primary vaccination series, antibody remained detectable for 6 months in 80%--100% (median: 100%) of persons and for 12 months in 58%--100% (median: 70%) (*61,64--69,71,85,99--103*). Among successfully immunized hemodialysis patients whose antibody titers subsequently declined below protective levels, limited data indicate that virtually all respond to a booster dose (*75*).

Duration of Vaccine-Induced Immunity. Among persons with normal immune status who respond to the primary series of hepatitis B vaccine, protection against hepatitis B persists even when antibody titers become undetectable (*97*). However, among hemodialysis patients who respond to the vaccine, protection against hepatitis B is not maintained when antibody titers fall below protective levels. In the U.S. vaccine efficacy trial, three hemodialysis patients who responded to the primary vaccination series developed HBV infection (*62*). One had received a kidney transplant 6 months before onset of infection, and anti-HBs had declined to borderline protective levels in the other two persons. In all three patients, infection resolved.

Alternative Routes of Administration. Among adults with normal immune status, intradermal administration of low doses of hepatitis B vaccine results in lower seroconversion rates (55%--81%) (*104--106*), and no data exist on long-term protection from this route of administration. Among infants and children, intradermal vaccination results in poor immunogenicity. Data are insufficient to evaluate alternative routes (e.g., intradermal) for vaccination among hemodialysis patients.

Hepatitis C Virus Infection

Epidemiology

Incidence and Prevalence. Data are limited on incidence of HCV infection among chronic hemodialysis patients. During 1982--1997, the incidence of non-A, non-B hepatitis among patients reported to CDC's national surveillance system decreased from 1.7% to 0.2% (*18*). The validity of these rates is uncertain because of inherent difficulties in diagnosing non-A, non-B hepatitis and probable variability in the application of diagnostic criteria by different dialysis centers. However, the downward trend can partially be explained by a decline in the rate of transfusion-associated disease after 1985 (*107,108*).

Since 1990, limited data from U.S. studies using testing for antibody to HCV (anti-HCV) to evaluate the incidence of HCV infection have reported annual rates of 0.73%--3% among hemodialysis patients (*109,110*). None of the patients who seroconverted had received transfusions in the interim or were injecting-drug users.

During 1992--1999, national surveillance data indicated that the proportion of centers that tested patients for anti-HCV increased from 22% to 56% (*18*) (CDC, unpublished data, 2001). In 1999, nationwide prevalence of anti-HCV was 8.9%, with some centers reporting prevalences >40% (CDC, unpublished data, 2001). Other studies of hemodialysis patients in the United States have reported anti-HCV prevalences of 10%--36% among adults (*109,111,112*) and 18.5% among children (*113*).

Transmission. HCV is most efficiently transmitted by direct percutaneous exposure to infectious blood, and like HBV, the chronically infected person is central to the epidemiology of HCV transmission. Risk factors associated with HCV infection among hemodialysis patients include history of blood transfusions, the volume of blood transfused, and years on dialysis (*114*). The number of years on dialysis is the major risk factor independently associated with higher rates of HCV infection. As the time patients spent on dialysis increased, their prevalence of HCV infection increased from an average of 12% for patients receiving dialysis <5 years to an average of 37% for patients receiving dialysis ≥5 years (*109,112,115*).

These studies, as well as investigations of dialysis-associated outbreaks of hepatitis C, indicate that HCV transmission most likely occurs because of inadequate infection control practices. During 1999--2000, CDC investigated three outbreaks of HCV infection among patients in chronic hemodialysis centers (CDC, unpublished data, 1999 and 2000). In two of the outbreaks, multiple transmissions of HCV occurred during periods of 16--24 months (attack rates: 6.6%--17.5%), and seroconversions were associated with receiving dialysis immediately after a chronically infected patient. Multiple opportunities for cross-contamination among patients were observed, including a) equipment and supplies that were not disinfected between patient use; b) use of common medication carts to prepare and distribute medications at patients' stations; c) sharing of multiple dose medication vials, which were placed at patients' stations on top of hemodialysis machines; d) contaminated priming buckets that were not routinely changed or cleaned and disinfected between patients; e) machine surfaces that were not routinely cleaned and disinfected between patients; and f) blood spills that were not cleaned up promptly. In the third outbreak, multiple new infections clustered at one point in time (attack rate: 27%), suggesting a common exposure event. Although the specific results of this investigation are pending, multiple opportunities for cross-contamination from chronically infected patients also were observed in this unit. In particular, supply carts were moved from one station to another and contained both clean supplies and blood-contaminated items, including small biohazard containers, sharps disposal boxes, and used vacutainers containing patients' blood.

Clinical Features and Natural History

Clinical Features and Natural History

HCV causes both acute and chronic hepatitis. The incubation period ranges from 14--180 days (average: 6--7 weeks) (*116*). Persons with newly acquired (acute) HCV infection typically are either asymptomatic or have a mild clinical illness. The course of acute hepatitis C is variable, although elevations in serum alanine aminotransferase (ALT) levels, often in a fluctuating pattern, are the most characteristic feature. Fulminant hepatic failure after acute hepatitis C is rare.

Most (average: 94%) hemodialysis patients with newly acquired HCV infection have elevated serum ALT levels (*117--121*). Elevations in serum ALT levels often precede anti-HCV seroconversion. Among prospectively followed transfusion recipients who developed acute HCV infection, elevated ALT levels preceded anti-HCV seroconversion (as measured by second generation assays) in 59%, and anti-HCV was detectable in most patients (78%) within 5 weeks after their first ALT elevation (*122*). However, elevations in ALT or aspartate aminotransferase (AST) levels can occur that are not related to viral hepatitis, and compared with ALT, AST is a less specific indicator of HCV-related liver disease among hemodialysis patients. In a recent outbreak investigation, only 28% of 25 hemodialysis patients with newly observed elevations in AST levels tested anti-HCV positive (CDC, unpublished data, 1999).

After acute HCV infection, 15%--25% of persons with normal immune status appear to resolve their infection without sequelae as defined by sustained absence of HCV RNA in serum and normalization of ALT (*123*). In some persons, ALT levels normalize, suggesting full recovery, but this is frequently followed by ALT elevations that indicate progression to chronic disease. Chronic HCV infection develops in most infected persons (75%--85%). Of persons with chronic HCV infection, 60%--70% have persistent or fluctuating ALT elevations, indicating active liver disease (*123*). Although similar rates of chronic liver disease have been observed among HCV-infected chronic hemodialysis patients (based on liver biopsy results), these patients might be less likely to have biochemical evidence of active liver disease (*124*). In seroprevalence studies of chronic hemodialysis patients, ALT elevations were reported in a median of 33.9% (range: 6%--73%) of patients who tested positive for anti-HCV (*117,124--136*).

No clinical or epidemiologic features among patients with acute infection have been reported to be predictive of either persistent infection or chronic liver disease. Most studies have reported that cirrhosis develops in 10%--20% of persons who have had chronic hepatitis C for 20--30 years, and hepatocellular carcinoma in 1%--5% (*123*). Extrahepatic manifestations of chronic HCV infection are considered to be of immunologic origin and include cryoglobulinemia, membranoproliferative glomerulonephritis, and porphyria cutanea tarda (*137*).

At least six different genotypes and >90 subtypes of HCV exist, with genotype 1 being the most common in the United States (*138,139*). Unlike HBV, infection with one HCV genotype or subtype does not protect against reinfection or superinfection with other HCV strains (*139*).

Alpha interferon alone or in combination with ribavirin is FDA-approved for the treatment of chronic hepatitis C (*48,140,141*). Combination therapy should be used with caution in patients with creatinine clearance <50 mL/minute and generally is contraindicated in patients with renal failure (*141,142*). Interferon monotherapy results in low sustained virologic response rates (*141,142*).

Screening and Diagnostic Tests

Serologic Assays. The only FDA-approved tests for diagnosis of HCV infection are those that measure anti-HCV and include enzyme immunoassays (EIAs) and a supplemental recombinant immunoblot assay (RIBA™) (*116*). These tests detect anti-HCV in ≥97% of infected persons, but do not distinguish between acute, chronic, or resolved infection. The average time from exposure to seroconversion is 8--9 weeks (*122*). Anti-HCV can be detected in 80% of patients within 15 weeks after exposure, in ≥90% within 5 months, and in ≥97% within 6 months (*122,143*). In rare instances, seroconversion can be delayed until 9 months after exposure (*143,144*). Anti-HCV persists indefinitely in most persons, but does not protect against reinfection.

As with any screening test, the positive predictive value of EIAs for anti-HCV is directly related to the prevalence of infection in the population and is low in populations with an HCV-infection prevalence <10% (*145,146*). Supplemental testing with a more specific assay (i.e., RIBA™) of a specimen with a positive anti-HCV result by EIA prevents reporting of false-positive results, particularly in settings where asymptomatic persons are being tested. Results of seroprevalence studies among chronic hemodialysis patients have indicated that 57%--100% of EIA positive results were RIBA™ positive (*124,126,128,133,135,147--152*), and 53%--100% were HCV RNA positive by reverse transcriptase polymerase chain reaction (RT-PCR) testing (*117,127,129,134,135*).

Nucleic Acid Detection. The diagnosis of HCV infection also can be made by qualitatively detecting HCV RNA using gene amplification techniques (e.g., RT-PCR) (*116*). HCV RNA can be detected in serum or plasma within 1--2 weeks after exposure and weeks before onset of ALT elevations or the appearance of anti-HCV. In rare instances, detection of HCV RNA might be the only evidence of HCV infection. Although a median of 3.4% (range: 0%--28%) of chronic hemodialysis patients who tested anti-HCV negative were HCV RNA positive, this might be an overestimate because follow-up samples to detect possible antibody seroconversions were not obtained on these patients (*117,118,126--128,130,131,133,134,148--154*).

Although not FDA-approved, RT-PCR assays for HCV infection are used commonly in clinical practice and are commercially available. Most RT-PCR assays have a lower limit of detection of 100--1,000 viral genome copies per mL. With adequate optimization of RT-PCR assays, 75%--85% of persons who are positive for anti-HCV and >95% of persons with acute or chronic hepatitis C will test positive for HCV RNA. Some HCV-infected persons might be only intermittently HCV RNA positive, particularly those with acute hepatitis C or with end-stage liver disease caused by hepatitis C. To minimize false-negative results, blood samples collected for RT-PCR should not contain heparin, and serum must be separated from cellular components within 2--4 hours after collection and preferably stored frozen at -20 C or -70 C (*155*). If shipping is required, frozen samples should be protected from thawing. Because of assay variability, rigorous quality assurance and control should be in place in clinical laboratories performing this assay, and proficiency testing is recommended.

Quantitative assays for measuring the concentration (i.e., titer) of HCV RNA have been developed and are available from commercial laboratories (*156*). These assays also are not FDA-approved and are less sensitive than qualitative RT-PCR assays (*157*). Quantitative assays should not be used as a primary test to confirm or exclude the diagnosis of HCV infection or to monitor the endpoint of treatment, and sequential measurement of HCV RNA levels has not proven useful in managing patients with hepatitis C.

Other Bloodborne Viruses

Hepatitis Delta Virus Infection

Delta hepatitis is caused by the hepatitis delta virus (HDV), a defective virus that causes infection only in persons with active HBV infection. The prevalence of HDV infection is low in the United States, with rates of <1%

among HBsAg-positive persons in the general population and ≥10% among HBsAg-positive persons with repeated percutaneous exposures (e.g., injecting-drug users, persons with hemophilia) (*158*). Areas of the world with high endemic rates of HDV infection include southern Italy, parts of Africa, and the Amazon Basin.

Few data exist on the prevalence of HDV infection among chronic hemodialysis patients, and only one transmission of HDV between such patients has been reported in the United States (*159*). In this episode, transmission occurred from a patient who was chronically infected with HBV and HDV to an HBsAg-positive patient after a massive bleeding incident; both patients received dialysis at the same station.

HDV infection occurs either as a co-infection with HBV or as a superinfection in a person with chronic HBV infection. Co-infection usually resolves, but superinfection frequently results in chronic HDV infection and severe disease. High mortality rates are associated with both types of infection. A serologic test that measures total antibody to HDV (anti-HDV) is commercially available.

Human Immunodeficiency Virus Infection

During 1985--1999, the percentage of U.S. hemodialysis centers that reported providing chronic hemodialysis for patients with HIV infection increased from 11% to 39%, and the proportion of hemodialysis patients with known HIV infection increased from 0.3% to 1.4% (*18*) (CDC, unpublished data, 2001).

HIV is transmitted by blood and other body fluids that contain blood. No patient-to-patient transmission of HIV has been reported in U.S. hemodialysis centers. However, such transmission has been reported in other countries; in one case, HIV transmission was attributed to mixing of reused access needles and inadequate disinfection of equipment (*160*).

HIV infection is usually diagnosed with assays that measure antibody to HIV, and a repeatedly positive EIA test should be confirmed by Western blot or another confirmatory test. Antiretroviral therapies for HIV-infected hemodialysis patients are commonly used and appear to be improving survival rates among this population. However, hepatotoxicity associated with certain protease inhibitors might limit the use of these drugs, especially in patients with underlying liver dysfunction (*161*).

Bacterial Infections

Epidemiology

Disease Burden. The annual mortality rate among hemodialysis patients is 23%, and infections are the second most common cause, accounting for 15% of deaths (*1*). Septicemia (10.9% of all deaths) is the most common infectious cause of mortality. In various studies evaluating rates of bacterial infections in hemodialysis outpatients, bacteremia occurred in 0.63%--1.7% of patients per month and vascular access infections (with or without bacteremia) in 1.3%--7.2% of patients per month (*162--170*). National surveillance data indicated that 4%--5% of patients received intravenous vancomycin during a 1-month period (and additional patients received other antimicrobials) (*18*). Although data on vancomycin use can be used to derive an estimate of the prevalence of suspected infections, the proportion of patients receiving antimicrobials who would fit a formal case definition for bacterial infection is unknown.

Infection Sites. In a study of 27 French hemodialysis centers, 28% of 230 infections in hemodialysis patients involved the vascular access, whereas 25% involved the lung, 23% the urinary tract, 9% the skin and soft tissues, and 15% other or unknown sites (*165*). Thirty-three percent of infections involved either the vascular access site or were bacteremias of unknown origin, many of which might have been caused by occult access infections. Thus, the vascular access site was the most common site for infection, but accounted for only one-third of infections. However, access site infections are particularly important because they can cause disseminated bacteremia or loss of the vascular access.

Vascular Access Infections. Vascular access infections are caused (in descending order of frequency) by *S. aureus,* coagulase-negative staphylococci (CNS), gram-negative bacilli, nonstaphylococcal gram-positive cocci (including enterococci), and fungi (*171*). The proportion of infections caused by CNS is higher among patients dialyzed through catheters than among patients dialyzed through fistulas or grafts.

The primary risk factor for access infection is access type, with catheters having the highest risk for infection, grafts intermediate, and native arteriovenous (AV) fistulas the lowest (*168*). Other potential risk factors for vascular access infections include a) location of the access in the lower extremity; b) recent access surgery; c) trauma, hematoma, dermatitis, or scratching over the access site; d) poor patient hygiene; e) poor needle insertion technique; f) older age; g) diabetes; h) immunosuppression; and i) iron overload (*164,167,172--175*).

Transmission. Bacterial pathogens causing infection can be either exogenous (i.e, acquired from contaminated dialysis fluids or equipment) or endogenous (i.e., caused by invasion of bacteria present in or on the patient). Exogenous pathogens have caused numerous outbreaks, most of which resulted from inadequate dialyzer reprocessing procedures (e.g., contaminated water or inadequate disinfectant) or inadequate treatment of municipal water for use in dialysis. During 1995--1997, four outbreaks were traced to contamination of the waste drain port on one type of dialysis machine (*176*). Recommendations to prevent such outbreaks are published elsewhere (*171*).

Contaminated medication vials also are a potential source of bacterial infection for patients. In 1999, an outbreak of *Serratia liquefaciens* bloodstream infections and pyrogenic reactions among hemodialysis patients was traced to contamination of vials of erythropoietin. These vials, which were intended for single use, were contaminated by repeated puncture to obtain additional doses and by pooling of residual medication into a common vial (*177*).

Endogenous pathogens first colonize the patient and later cause infection. Colonization means that microorganisms have become resident in or on the body (e.g., in the nares or stool); a culture from the site is positive, but no symptoms or signs of infection exist. Colonization with potentially pathogenic microorganisms, often unknown to staff members, is common in patients with frequent exposure to hospitals and other health-care settings. Colonization most often occurs when microorganisms are transmitted from a colonized or infected source patient to another patient on the hands of health-care workers who do not comply with infection control precautions. Less commonly, contamination of environmental surfaces (e.g., bed rails, countertops) plays a role (*178*).

Infection occurs when microorganisms invade the body, damaging tissue and causing signs or symptoms of infection, and is aided by invasive devices (e.g., the hemodialysis vascular access). Evidence exists that when prevalence of colonization in a population is less frequent, infection in that population will also be less frequent, and infection control recommendations for hemodialysis units are designed to prevent colonization (*179*). Additional measures designed to prevent infection from colonizing organisms (e.g., using aseptic technique during vascular access) are presented elsewhere (*180*).

Antimicrobial Resistance

Antimicrobial-resistant bacteria are more common in patients with severe illness, who often have had multiple hospitalizations or surgical procedures, and in those who have received prolonged courses of antimicrobial agents. In health-care settings, including hemodialysis centers, such patients can serve as a source for transmission.

Clinically important drug-resistant bacteria that commonly cause health-care--associated infections include MRSA, methicillin-resistant CNS, VRE, and multidrug-resistant gram negative rods, including strains of *Pseudomonas aeruginosa*, *Stenotrophomonas maltophilia*, and *Acinetobacter* species, some of which are resistant to all available antimicrobials. In addition, strains of *S. aureus* with intermediate resistance to vancomycin and other glycopeptide antibiotics have recently been reported; these strains are called vancomycin-intermediate *S. aureus* (VISA) or glycopeptide-intermediate *S. aureus* (GISA) (*181,182*). Intermediate resistance to vancomycin is reported even more frequently among CNS (*183,184*).

Hemodialysis patients have played a prominent role in the epidemic of vancomycin resistance. In 1988, a renal unit in London, England, reported one of the first cases of VRE (*185*). In three studies, 12%--22% of hospitalized patients infected or colonized with VRE were receiving hemodialysis (*178,186,187*). Furthermore, three of the first five patients identified with VISA (or GISA) were on chronic hemodialysis, and one had received acute dialysis (*182*).

Prevalence of VRE has increased rapidly at U.S. hospitals; among intensive care unit patients with nosocomial infections reported to the National Nosocomial Infections Surveillance (NNIS) system, the percentage of enterococcal isolates resistant to vancomycin increased from 0.5% in 1989 to 25.2% in 1999 (*23*) (CDC, unpublished data, 2000). This increase is attributable to patient-to-patient transmission in health-care settings and transmission of resistant genes among previously susceptible enterococci. Once vancomycin resistance has been transferred to a patient, antimicrobials select for resistant organisms, causing them to increase in number relative to susceptible organisms. Prevalence of VRE colonization among patients varies in different health-care settings; in hemodialysis centers, the reported prevalence in stool samples ranged from 1% to 9% (*188,189*). In one center with a prevalence of 9%, three patients developed VRE infections in 1 year (*188*).

Vancomycin Use

Dialysis patients have played a prominent role in the epidemic of vancomycin resistance because this drug is used commonly in these patients, in part because vancomycin can be conveniently administered to patients when they come in for hemodialysis treatments. However, two studies indicate that cefazolin, a first-generation cephalosporin, could be substituted for vancomycin in many patients (*190,191*). One of these studies reported that many pathogens causing infections in hemodialysis patients are susceptible to cefazolin (*190*), and both studies reported therapeutic cefazolin blood levels 48--72 hours after dosing, making in-center administration three times a week after dialysis feasible.

Equipment, Supplies, and Environmental Surfaces

The hemodialysis machine and its components also can be vehicles for patient-to-patient transmission of bloodborne viruses and pathogenic bacteria (*24,192*). The external surfaces of the machine are the most likely sources for contamination. These include not only frequently touched surfaces (e.g., the control panel), but also attached waste containers used during the priming of the dialyzers, blood tubing draped or clipped to waste containers, and items placed on tops of machines for convenience (e.g., dialyzer caps and medication vials).

Sterilization, Disinfection, and Cleaning

A sterilization procedure kills all microorganisms, including highly resistant bacterial spores (*24*). Sterilization procedures are most commonly accomplished by steam or ethylene oxide gas. For products that are heat sensitive, an FDA-cleared liquid chemical sterilant can be used with a long exposure time (i.e., 3--10 hours).

High-level disinfection kills all viruses and bacteria, but not high numbers of bacterial spores. High-level disinfection can be accomplished by heat pasteurization or, more commonly, by an FDA-cleared chemical sterilant, with an exposure time of 12--45 minutes. Sterilants and high-level disinfectants are designed to be used on medical devices, not environmental surfaces. Intermediate-level disinfection kills bacteria and most viruses and is accomplished by using a tuberculocidal "hospital disinfectant" (a term used by the U.S. Environmental Protection Agency [EPA] in registering germicides) or a 1:100 dilution of bleach (300--600 mg/L free chlorine). Low-level disinfection kills most bacteria and is accomplished by using general purpose disinfectants. Intermediate and low-level disinfectants are designed to be used on environmental surfaces; they also can be used on noncritical medical devices, depending on the design and labeling claim.

Cleaning eliminates dirt and some bacteria and viruses and is accomplished by using a detergent or detergent germicide. Antiseptics (e.g., formulations with povidone-iodine, hexachlorophene, or chlorhexidene) are designed for use on skin and tissue and should not be used on medical equipment or environmental surfaces.

Regardless of the procedure used, cleaning with a germicidal detergent before disinfection (or sterilization) is essential to remove organic material (e.g., blood, mucous, or feces), dirt, or debris. The presence of such material protects microorganisms from the sterilization or disinfection process by physically blocking or inactivating the disinfectant or sterilant.

The choice of what procedure or which chemical germicide to use for medical devices, instruments, and environmental surfaces depends on several factors, including the need to maintain the structural integrity and function of the item and how the item will be used. Three general categories of use for medical items are recognized, each of which require different levels of sterilization or disinfection (*193*). These categories are a) critical, which includes items introduced directly into the bloodstream or normally sterile areas of the body (e.g., needles, catheters, hemodialyzers, blood tubing); b) semicritical, which includes equipment that comes in contact with intact mucous membranes (e.g., fiberoptic endoscopes, glass thermometers); and c) noncritical, which includes equipment that touches only intact skin (e.g., blood pressure cuffs). Semicritical items are not generally used in dialysis units.

Internal Pathways of Hemodialysis Machines. In single-pass hemodialysis machines, the internal fluid pathways are not subject to contamination with blood. If a dialyzer leak occurs, dialysis fluid might become contaminated with blood, but this contaminated fluid is discarded through a drain and does not return to the dialysis machine to contaminate predialyzer surfaces. For dialysis machines that use a dialysate recirculating system (e.g., some ultrafiltration control machines and those that regenerate the dialysate), a blood leak in a dialyzer could contaminate the internal pathways of the machine, which could in turn contaminate the dialysis fluid of subsequent patients (*192*). However, procedures normally practiced after each use (i.e., draining the dialysis fluid and rinsing and disinfecting the machine) will reduce the level of contamination to below infectious levels. In addition, an intact dialyzer membrane will not allow passage of bacteria or viruses (*24*).

Pressure transducer filter protectors are used primarily to prevent contamination and preserve the functioning of the pressure monitoring (i.e., arterial, venous, or both) components of the hemodialysis machine. Hemodialysis machines usually have both external (typically supplied with the blood tubing set) and internal protectors, with the internal protector serving as a backup in case the external transducer protector fails. Failure to use an external protector or to replace the protector when it becomes contaminated (i.e., wetted with saline or blood) can result in contamination of the internal transducer protector, which in turn could allow transmission of bloodborne pathogens (*24*). However, no epidemiologic evidence exists that contamination of the internal transducer protector caused by failure of the external transducer protector has led to either mixing of blood or the transmission of bloodborne agents.

Dialyzer Reprocessing. Approximately 80% of U.S. chronic hemodialysis centers reprocess (i.e., reuse) dialyzers for the same patient (*18*), and guidelines for reprocessing have been published elsewhere (see Suggested Readings). Although outbreaks of bacterial infections and pyrogenic reactions have occurred because of inadequate reprocessing procedures and failure to maintain standards for water quality, reuse has not been associated with transmission of bloodborne viruses. Any theoretical risk for HBV transmission from reuse of dialyzers would primarily affect staff members who handle these dialyzers. Although no increase in HBV (or HCV) infection among staff members who work in such centers has been reported, many centers do not reuse dialyzers from HBsAg-positive patients (*24*).

Infection Control Precautions for Outpatient Hemodialysis Settings Compared with Inpatient Hospital Settings

Contact transmission is the most important route by which pathogens are transmitted in health-care settings, including hemodialysis units. Contact transmission occurs most commonly when microorganisms from a patient are transferred to the hands of a health-care worker who does not comply with infection control precautions, then touches another patient. Less commonly, environmental surfaces (e.g., bed rails, countertops) become contaminated and serve as an intermediate reservoir for pathogens; transmission can occur when a worker touches the surface then touches a patient or when a patient touches the surface.

In the hemodialysis setting, contact transmission plays a major role in transmission of bloodborne pathogens. If a health-care worker's hands become contaminated with virus-infected blood from one patient, the worker can transfer the virus to a second patient's skin or blood line access port, and the virus can be inoculated into that patient when the skin or access port is punctured with a needle.

Contact transmission can be prevented by hand hygiene (i.e., hand washing or use of a waterless hand rub), glove use, and disinfection of environmental surfaces. Of these, hand hygiene is the most important. In addition, nonsterile disposable gloves provide a protective barrier for workers' hands, preventing them from becoming soiled or contaminated, and reduce the likelihood that microorganisms present on the hands of personnel will be transmitted to patients. However, even with glove use, hand washing is needed because pathogens deposited on the outer surface of gloves can be detected on hands after glove removal, possibly because of holes or defects in the gloves, leakage at the wrist, or contamination of hands during glove removal (*194*).

Standard Precautions are the system of infection control precautions recommended for the inpatient hospital setting (*195*). Standard Precautions are used on all patients and include use of gloves, gown, or mask whenever needed to prevent contact of the health-care worker with blood, secretions, excretions, or contaminated items.

In addition to Standard Precautions, more stringent precautions are recommended for hemodialysis units because of the increased potential for contamination with blood and pathogenic microorganisms (see Infection Control Practices Recommended for Hemodialysis Units). For example, infection control practices for hemodialysis units restrict the use of common supplies, instruments, medications, and medication trays and prohibit the use of a common medication cart.

For certain patients, including those infected or colonized with MRSA or VRE, contact precautions are used in the inpatient hospital setting. Contact precautions include a) placing the patient in a single room or with another patient infected or colonized with the same organism; b) using gloves whenever entering the patient's room; and c) using a gown when entering the patient's room if the potential exists for the worker's clothing to have substantial contact with the patient, environmental surfaces, or items in the patient's room. Workers also should wear a gown if the patient has diarrhea, an ileostomy, a colostomy, or wound drainage not contained by a dressing.

However, contact precautions are not recommended in hemodialysis units for patients infected or colonized with pathogenic bacteria for several reasons. First, although contact transmission of pathogenic bacteria is well-documented in hospitals, similar transmission has not been well-documented in hemodialysis centers. Transmission might not be apparent in dialysis centers, possibly because it occurs less frequently than in acute-care hospitals or results in undetected colonization rather than overt infection. Also, because dialysis patients are frequently hospitalized, determining whether transmission occurred in the inpatient or outpatient setting is difficult. Second, contamination of the patient's skin, bedclothes, and environmental surfaces with pathogenic bacteria is likely to be more common in hospital settings (where patients spend 24 hours a day) than in outpatient hemodialysis centers (where patients spend approximately 10 hours a week). Third, the routine use of infection control practices recommended for hemodialysis units, which are more stringent than the Standard Precautions routinely used in hospitals, should prevent transmission by the contact route.

RECOMMENDATIONS

Rationale

Preventing transmission among chronic hemodialysis patients of bloodborne viruses and pathogenic bacteria from both recognized and unrecognized sources of infection requires implementation of a comprehensive infection control program. The components of such a program include infection control practices specifically designed for the hemodialysis setting, including routine serologic testing and immunization, surveillance, and training and education (Box).

The infection control practices recommended for hemodialysis units will reduce opportunities for patient-to-patient transmission of infectious agents, directly or indirectly via contaminated devices, equipment and supplies, environmental surfaces, or hands of personnel. These practices should be carried out routinely for all patients in the chronic hemodialysis setting because of the increased potential for blood contamination during hemodialysis and because many patients are colonized or infected with pathogenic bacteria. Such practices include additional measures to prevent HBV transmission because of the high titer of HBV and its ability to survive on environmental surfaces. For patients at increased risk for transmission of pathogenic bacteria, including antimicrobial-resistant strains, additional precautions also might be necessary in some circumstances. Furthermore, surveillance for infections and other adverse events is required to monitor the effectiveness of infection control practices, as well as training and education of both staff members and patients to ensure that appropriate infection control behaviors and techniques are carried out.

Infection Control Practices for Hemodialysis Units

In each chronic hemodialysis unit, policies and practices should be reviewed and updated to ensure that infection control practices recommended for hemodialysis units are implemented and rigorously followed (see Recommended Infection Control Practices for Hemodialysis Units at a Glance). Intensive efforts must be made to educate new staff members and reeducate existing staff members regarding these practices.

Infection Control Precautions for All Patients

During the process of hemodialysis, exposure to blood and potentially contaminated items can be routinely anticipated; thus, gloves are required whenever caring for a patient or touching the patient's equipment. To facilitate glove use, a supply of clean nonsterile gloves and a glove discard container should be placed near each dialysis station. Hands always should be washed after gloves are removed and between patient contacts, as well as after touching blood, body fluids, secretions, excretions, and contaminated items. A sufficient number of sinks with warm water and soap should be available to facilitate hand washing. If hands are not visibly soiled, use of a waterless antiseptic hand rub can be substituted for hand washing.

Any item taken to a patient's dialysis station could become contaminated with blood and other body fluids and serve as a vehicle of transmission to other patients either directly or by contamination of the hands of personnel. Therefore, items taken to a patient's dialysis station, including those placed on top of dialysis machines, should either be disposed of, dedicated for use only on a single patient, or cleaned and disinfected before being returned to a common clean area or used for other patients. Unused medications or supplies (e.g., syringes, alcohol swabs) taken to the patient's station should not be returned to a common clean area or used on other patients.

Additional measures to prevent contamination of clean or sterile items include a) preparing medications in a room or area separated from the patient treatment area and designated only for medications; b) not handling or storing contaminated (i.e., used) supplies, equipment, blood samples, or biohazard containers in areas where medications and clean (i.e., unused) equipment and supplies are handled; and c) delivering medications separately to each patient. Common carts should not be used within the patient treatment area to prepare or distribute medications. If trays are used to distribute medications, clean them before using for a different patient.

Intravenous medication vials labeled for single use, including erythropoetin, should not be punctured more than once (*196,197*). Once a needle has entered a vial labeled for single use, the sterility of the product can no longer be guaranteed. Residual medication from two or more vials should not be pooled into a single vial.

If a common supply cart is used to store clean supplies in the patient treatment area, this cart should remain in a designated area at a sufficient distance from patient stations to avoid contamination with blood. Such carts should not be moved between stations to distribute supplies.

Staff members should wear gowns, face shields, eye wear, or masks to protect themselves and prevent soiling of clothing when performing procedures during which spurting or spattering of blood might occur (e.g., during initiation and termination of dialysis, cleaning of dialyzers, and centrifugation of blood). Such protective clothing or gear should be changed if it becomes soiled with blood, body fluids, secretions, or excretions. Staff members should not eat, drink, or smoke in the dialysis treatment area or in the laboratory. However, patients can be served meals or eat food brought from home at their dialysis station. The glasses, dishes, and other utensils should be cleaned in the usual manner; no special care of these items is needed.

Cleaning and Disinfection. Establish written protocols for cleaning and disinfecting surfaces and equipment in the dialysis unit, including careful mechanical cleaning before any disinfection process (Table 2). If the manufacturer has provided instructions on sterilization or disinfection of the item, these instructions should be followed. For each chemical sterilant and disinfectant, follow the manufacturer's instructions regarding use, including appropriate dilution and contact time.

After each patient treatment, clean environmental surfaces at the dialysis station, including the dialysis bed or chair, countertops, and external surfaces of the dialysis machine, including containers associated with the prime waste. Use any soap, detergent, or detergent germicide. Between uses of medical equipment (e.g., scissors, hemostats, clamps, stethoscopes, blood pressure cuffs), clean and apply a hospital disinfectant (i.e., low-level disinfection); if the item is visibly contaminated with blood, use a tuberculocidal disinfectant (i.e., intermediate-level disinfection).

For a blood spill, immediately clean the area with a cloth soaked with a tuberculocidal disinfectant or a 1:100 dilution of household bleach (300--600 mg/L free chlorine) (i.e., intermediate-level disinfection). The staff member doing the cleaning should wear gloves, and the cloth should be placed in a bucket or other leakproof container. After all visible blood is cleaned, use a new cloth or towel to apply disinfectant a second time.

Published methods should be used to clean and disinfect the water treatment and distribution system and the internal circuits of the dialysis machine, as well as to reprocess dialyzers for reuse (see Suggested Readings). These methods are designed to control bacterial contamination, but will also eliminate bloodborne viruses. For single-pass machines, perform rinsing and disinfection procedures at the beginning or end of the day. For batch recirculating machines, drain, rinse, and disinfect after each use. Follow the same methods for cleaning and disinfection if a blood leak has occurred, regardless of the type of dialysis machine used. Routine bacteriologic assays of water and dialysis fluids should be performed according to the recommendations of the Association for the Advancement of Medical Instrumentation (see Suggested Readings).

Venous pressure transducer protectors should be used to cover pressure monitors and should be changed between patients, not reused. If the external transducer protector becomes wet, replace immediately and inspect the protector. If fluid is visible on the side of the transducer protector that faces the machine, have qualified personnel open the machine after the treatment is completed and check for contamination. This includes inspection for possible blood contamination of the internal pressure tubing set and pressure sensing port. If contamination has occurred, the machine must be taken out of service and disinfected using either 1:100 dilution of bleach (300--600 mg/L free chlorine) or a commercially available, EPA-registered tuberculocidal germicide before reuse. Frequent blood line pressure alarms or frequent adjusting of blood drip chamber levels can be an indicator of this problem. Taken separately, these incidents could be characterized as isolated malfunctions. However, the potential public health significance of the total number of incidents nationwide make it imperative that all incidents of equipment contamination be reported immediately to the FDA (800-FDA-1088).

Housekeeping staff members in the dialysis facility should promptly remove soil and potentially infectious waste and maintain an environment that enhances patient care. All disposable items should be placed in bags thick enough to prevent leakage. Wastes generated by the hemodialysis facility might be contaminated with blood and should be considered infectious and handled accordingly. These solid medical wastes should be disposed of properly in an incinerator or sanitary landfill, according to local and state regulations governing medical waste disposal.

Hemodialysis in Acute-Care Settings. For patients with acute renal failure who receive hemodialysis in acute-care settings, Standard Precautions as applied in all health-care settings are sufficient to prevent transmission of bloodborne viruses. However, when chronic hemodialysis patients receive maintenance hemodialysis while hospitalized, infection control precautions specifically designed for chronic hemodialysis units (see

Recommended Practices at a Glance) should be applied to these patients. If both acute and chronic renal failure patients receive hemodialysis in the same unit, these infection control precautions should be applied to all patients.

Regardless of where in the acute-care setting chronic hemodialysis patients receive dialysis, the HBsAg status of all such patients should be ascertained at the time of admission to the hospital, by either a written report from the referring center (including the most recent date testing was performed) or by a serologic test. The HBV serologic status should be prominently placed in patients' hospital records, and all health-care personnel assigned to these patients, as well as the infection control practitioner, should be aware of the patients' serologic status. While hospitalized, HBsAg-positive chronic hemodialysis patients should undergo dialysis in a separate room and use separate machines, equipment, instruments, supplies, and medications designated only for HBsAg-positive patients (see Prevention and Management of HBV Infection). While HBsAg-positive patients are receiving dialysis, staff members who are caring for them should not care for susceptible patients.

Routine Serologic Testing

Chronic Hemodialysis Patients. Routinely test all chronic hemodialysis patients for HBV and HCV infection (see Recommended Practices at a Glance), promptly review results, and ensure that patients are managed appropriately based on their testing results (see later recommendations for each virus). Communicate test results (positive and negative) to other units or hospitals when patients are transferred for care. Routine testing for HDV or HIV infection for purposes of infection control is not recommended.

The HBV serologic status (i.e., HBsAg, total anti-HBc, and anti-HBs) of all patients should be known before admission to the hemodialysis unit. For patients transferred from another unit, test results should be obtained before the patients' transfer. If a patient's HBV serologic status is not known at the time of admission, testing should be completed within 7 days. The hemodialysis unit should ensure that the laboratory performing the testing for anti-HBs can define a 10 mIU/mL concentration to determine protective levels of antibody.

Routine HCV testing should include use of both an EIA to test for anti-HCV and supplemental or confirmatory testing with an additional, more specific assay (Figure). Use of RT-PCR for HCV RNA as the primary test for routine screening is not recommended because few HCV infections will be identified in anti-HCV negative patients. However, if ALT levels are persistently abnormal in patients who are anti-HCV negative in the absence of another etiology, testing for HCV RNA should be considered (for proper specimen collection and handling, see Hepatitis C Virus Infection, Screening and Diagnostic Tests).

Hemodialysis Staff Members. Previously, testing for HBV infection was recommended for all staff members at the time of employment and for susceptible staff members at routine intervals thereafter (*198*); however, such testing is no longer considered necessary. The risk for HBV infection among hemodialysis staff members is no greater than that for other health-care workers. Thus, routine testing of staff members is not recommended except when required to document response to hepatitis B vaccination (see Postvaccination Testing and Revaccination of Nonresponders). Routine testing of staff members for HCV, HDV, or HIV infection is not recommended.

Hepatitis B Vaccination

Vaccine Schedule and Dose. Hepatitis B vaccination is recommended for all susceptible chronic hemodialysis patients and for all staff members (Table 3). Vaccination is recommended for pre--end-stage renal disease patients before they become dialysis dependent and for peritoneal and home dialysis patients because they might require in-center hemodialysis. Hepatitis B vaccine should be administered by the intramuscular route and only in the deltoid muscle for adults and children. Intradermal or subcutaneous administration of hepatitis B vaccine is not recommended.

If an adult patient begins the vaccine series with a standard dose before beginning hemodialysis treatment, then moves to hemodialysis treatment before completing the series, complete the series using the higher dose recommended for hemodialysis patients (Table 3). No specific recommendations have been made for higher doses for pediatric hemodialysis patients. If a lower than recommended vaccine dose is administered to either adults or children, the dose should be repeated.

If the vaccination series is interrupted after the first dose, the second dose should be administered as soon as possible. For the three-dose primary vaccine series, the second and third doses should be separated by an interval of at least 2 months; if only the third dose is delayed, that dose should be administered when convenient. When hepatitis B vaccine has been administered at the same time as other vaccines, no interference with the antibody response of the other vaccines has been demonstrated.

Postvaccination Testing and Revaccination of Nonresponders. Test all vaccinees for anti-HBs 1--2 months after the last primary vaccine dose, to determine their response to the vaccine (adequate response is defined as ≥10 mIU/mL). Patients and staff members who do not respond to the primary vaccine series should be revaccinated with three additional doses and retested for response. No additional doses of vaccine are warranted for those who do not respond to the second series.

Evaluate staff members who do not respond to revaccination to determine if they are HBsAg positive (*199*). Persons who are HBsAg positive should be counseled accordingly (e.g., need for medical evaluation, vaccination of sexual and household contacts). Primary nonresponders to vaccination who are HBsAg negative should be considered susceptible to HBV infection and counseled regarding precautions to prevent HBV infection and the need to obtain postexposure prophylaxis with hepatitis B immune globulin for any known or probable percutaneous or mucosal exposure to HBsAg-positive blood (*199*).

Follow-Up of Vaccine Responders. Retest patients who respond to the vaccine annually for anti-HBs. If anti-HBs declines to <10 mIU/mL, administer a booster dose of hepatitis B vaccine and continue to retest annually. Retesting immediately after the booster dose is not necessary. For staff members who respond to the vaccine, booster doses of vaccine are not necessary, and periodic serologic testing to monitor antibody concentrations is not recommended (*199*).

Patients with a History of Vaccination. Routine childhood vaccination against hepatitis B has been recommended since 1991 and routine adolescent vaccination since 1995 (*89,198*). Thus, many persons who develop end-stage renal failure will have a history of vaccination against hepatitis B. These persons should have responded to the vaccine when their immune status was normal, but if their anti-HBs levels are <10 mIU/mL when they begin dialysis, they should be revaccinated with a complete primary series.

Prevention and Management of HBV Infection

Preventing HBV transmission among chronic hemodialysis patients requires a) infection control precautions recommended for all hemodialysis patients; b) routine serologic testing for markers of HBV infection and prompt review of results; c) isolation of HBsAg-positive patients with dedicated room, machine, other equipment, supplies, and staff members; and d) vaccination. Additional infection control practices are needed because of the potential for environmentally mediated transmission of HBV, rather than internal contamination of dialysis machines. The need for routine follow-up testing, vaccination, or isolation is based on patients' serologic status (Table 1 and Recommended Practices at a Glance).

HBV-Susceptible Patients. Vaccinate all susceptible patients (see Hepatitis B Vaccination). Test susceptible patients monthly for HBsAg, including those who a) have not yet received hepatitis B vaccine, b) are in the process of being vaccinated, or c) have not adequately responded to vaccination. Although the incidence of HBV infection is low among chronic hemodialysis patients, preventing transmission depends on timely detection of patients converting from HBsAg negative to HBsAg positive and rapid implementation of isolation procedures before cross-contamination can occur.

HBsAg Seroconversions. Report HBsAg-positive seroconversions to the local health department as required by law or regulation. When a seroconversion occurs, review all patients' routine laboratory test results to identify additional cases. Perform additional testing as indicated later in this section. Investigate potential sources for infection to determine if transmission might have occurred within the dialysis unit, including review of newly infected patients' recent medical history (e.g., blood transfusion, hospitalization), history of high-risk behavior (e.g., injecting-drug use, sexual activity), and unit practices and procedures.

In patients newly infected with HBV, HBsAg often is the only serologic marker initially detected; repeat HBsAg testing and test for anti-HBc (including IgM anti-HBc) 1--2 months later. Six months later, repeat HBsAg testing and test for anti-HBs to determine clinical outcome and need for counseling, medical evaluation, and vaccination of contacts. Patients who become HBsAg negative are no longer infectious and can be removed from isolation.

HBV-Infected Patients. To isolate HBsAg-positive patients, designate a separate room for their treatment and dedicate machines, equipment, instruments, supplies, and medications that will not be used by HBV-susceptible patients. Most importantly, staff members who are caring for HBsAg-positive patients should not care for susceptible patients at the same time, including during the period when dialysis is terminated on one patient and initiated on another.

Newly opened units should have isolation rooms for the dialysis of HBsAg-positive patients. For existing units in which a separate room is not possible, HBsAg-positive patients should be separated from HBV-susceptible patients in an area removed from the mainstream of activity and should undergo dialysis on dedicated machines. If a machine that has been used on an HBsAg-positive patient is needed for an HBV-susceptible patient, internal pathways of the machine can be disinfected using conventional protocols and external surfaces cleaned using soap and water or a detergent germicide.

Dialyzers should not be reused on HBsAg-positive patients. Because HBV is efficiently transmitted through occupational exposure to blood, reprocessing dialyzers from HBsAg-positive patients might place HBV-susceptible staff members at increased risk for infection.

Chronically infected patients (i.e., those who are HBsAg positive, total anti-HBc positive, and IgM anti-HBc negative) are infectious to others and are at risk for chronic liver disease. They should be counseled regarding preventing transmission to others, their household and sexual partners should receive hepatitis B vaccine, and they should be evaluated (by consultation or referral, if appropriate) for the presence or development of chronic liver disease according to current medical practice guidelines. Persons with chronic liver disease should be vaccinated against hepatitis A, if susceptible.

Chronically infected patients do not require any routine follow-up testing for purposes of infection control. However, annual testing for HBsAg is reasonable to detect the small percentage of HBV-infected patients who might lose their HBsAg.

HBV-Immune Patients. Annual anti-HBs testing of patients who are positive for anti-HBs ($\geq$10 mIU/mL) and negative for anti-HBc determines the need for booster doses of vaccine to ensure that protective levels of antibody are maintained. No routine follow-up testing is necessary for patients who are positive for both anti-HBs and anti-HBc.

HBV-immune patients can undergo dialysis in the same area as HBsAg-positive patients, or they can serve as a geographic buffer between HBsAg-positive and HBV-susceptible patients. Staff members can be assigned to care for both infected and immune patients on the same shift.

Isolated Anti-HBc--Positive Patients. Patients who test positive for isolated anti-HBc (i.e., those who are anti-HBc positive, HBsAg negative, and anti-HBs negative) should be retested on a separate serum sample for total anti-HBc, and if positive, for IgM anti-HBc. The following guidelines should be used for interpretation and follow-up:

- If total anti-HBc is negative, consider patient susceptible, and follow recommendations for vaccination.
- If total anti-HBc is positive and IgM anti-HBc is negative, follow recommendations for vaccination.
    - If anti-HBs is <10 mIU/mL even after revaccination, test for HBV DNA.

    - If HBV DNA is negative, consider patient susceptible (i.e., the anti-HBc result is a false positive), and test monthly for HBsAg.
    - If HBV DNA is positive, consider patient as having past infection or "low-level" chronic infection (i.e., the anti-HBc result is a true positive); no further testing is necessary.
    - Isolation is not necessary because HBsAg is not detectable.
- If both total and IgM anti-HBc are positive, consider patient recently infected and test for anti-HBs in 4--6 months; no further routine testing is necessary.
    - Isolation is not necessary because HBsAg is not detectable.

Prevention and Management of HCV Infection

HCV transmission within the dialysis environment can be prevented by strict adherence to infection control precautions recommended for all hemodialysis patients (see Recommended Practices at a Glance). Although isolation of HCV-infected patients is not recommended, routine testing for ALT and anti-HCV is important for monitoring transmission within centers and ensuring that appropriate precautions are being properly and consistently used.

HCV-Negative Patients. Monthly ALT testing will facilitate timely detection of new infections and provide a pattern from which to determine when exposure or infection might have occurred. In the absence of unexplained ALT elevations, testing for anti-HCV every 6 months should be sufficient to monitor the occurrence of new HCV infections. If unexplained ALT elevations are observed in patients who are anti-HCV negative, repeat anti-HCV testing is warranted. If unexplained ALT elevations persist in patients who repeatedly test anti-HCV negative, testing for HCV RNA should be considered.

Anti-HCV Seroconversions. Report anti-HCV--positive seroconversions to the local health department as required by law or regulation. When a seroconversion occurs, review all other patients' routine laboratory test results to identify additional cases. Perform additional testing as indicated later in this section. Investigate potential sources for infection to determine if transmission might have occurred within the dialysis unit, including review of newly infected patients' recent medical history (e.g., blood transfusion, hospitalization), history of high-risk behavior (e.g., injecting-drug use, sexual activity), and unit practices and procedures.

If ≥1 patient seroconverts from anti-HCV negative to positive during a 6-month period, more frequent (e.g., every 1--3 months) anti-HCV testing of HCV-negative patients could be warranted for a limited time (e.g., 3--6 months) to detect additional infections. If no additional newly infected patients are identified, resume semiannual testing. If ongoing HCV transmission among patients is identified, implement control measures based on results of investigation of potential sources for transmission and monitor their effectiveness (e.g., perform more frequent anti-HCV testing of HCV-negative patients for 6--12 months before resuming semiannual testing).

HCV-Positive Patients. Patients who are anti-HCV positive (or HCV RNA positive) do not have to be isolated from other patients or dialyzed separately on dedicated machines. Furthermore, they can participate in dialyzer reuse programs. Unlike HBV, HCV is not transmitted efficiently through occupational exposures. Thus, reprocessing dialyzers from HCV-positive patients should not place staff members at increased risk for infection.

HCV-positive persons should be evaluated (by consultation or referral, if appropriate) for the presence or development of chronic liver disease according to current medical practice guidelines. They also should receive information concerning how they can prevent further harm to their liver and prevent transmitting HCV to others (*116,141*). Persons with chronic liver disease should be vaccinated against hepatitis A, if susceptible.

Prevention and Management of HDV Infection

Because of the low prevalence of HDV infection in the United States, routine testing of hemodialysis patients is not necessary or recommended. However, if a patient is known to be infected with HDV, or if evidence exists of transmission of HDV in a dialysis center, screening for delta antibody is warranted. Because HDV depends on an HBV-infected host for replication, prevention of HBV infection will prevent HDV infection in a person susceptible to HBV. Patients who are known to be infected with HDV should be isolated from all other dialysis patients, especially those who are HBsAg-positive.

Prevention and Management of HIV Infection

Routine testing of hemodialysis patients for HIV infection for infection control purposes is not necessary or recommended. However, patients with risk factors for HIV infection should be tested so that, if infected, they can receive proper medical care and counseling regarding preventing transmission of the virus (*201*).

Infection control precautions recommended for all hemodialysis patients (see Recommended Practices at a Glance) are sufficient to prevent HIV transmission between patients. HIV-infected patients do not have to be isolated from other patients or dialyzed separately on dedicated machines. In addition, they can participate in dialyzer reuse programs. Because HIV is not transmitted efficiently through occupational exposures, reprocessing dialyzers from HIV-positive patients should not place staff members at increased risk for infection.

Prevention and Management of Bacterial Infections

Follow published guidelines for judicious use of antimicrobials, particularly vancomycin, to reduce selection for antimicrobial-resistant pathogens (*202*). Infection control precautions recommended for all hemodialysis patients (see Recommended Practices at a Glance) are adequate to prevent transmission for most patients infected or colonized with pathogenic bacteria, including antimicrobial-resistant strains. However, additional infection control precautions should be considered for treatment of patients who might be at increased risk for transmitting pathogenic bacteria. Such patients include those with either a) an infected skin wound with drainage that is not contained by dressings (the drainage does not have to be culture positive for VRE, MRSA, or any specific pathogen) or b) fecal incontinence or diarrhea uncontrolled with personal hygiene measures. For these patients, consider using the following additional precautions: a) staff members treating the patient should wear a separate gown over their usual clothing and remove the gown when finished caring for the patient and b) dialyze the patient at a station with as few adjacent stations as possible (e.g., at the end or corner of the unit).

SURVEILLANCE FOR INFECTIONS AND OTHER ADVERSE EVENTS

Develop and maintain a separate centralized record-keeping system (e.g., log book or electronic file) to record the results of patients' vaccination status, serologic testing results for viral hepatitis (including ALT), episodes of bacteremia or loss of the vascular access caused by infection (including date of onset, site of infection, genus and species of the infecting organism, and selected antimicrobial susceptibility results),* and adverse events (e.g., blood leaks and spills, dialysis machine malfunctions). Designate a staff person to promptly review the results of routine testing each time such testing is performed and periodically review recorded episodes of bacteremia or vascular access infections. Specify a procedure for actions required when changes occur in test results or in the frequency of episodes of bacteremias or vascular access loss because of infection. Maintain records for each patient that include the location of the dialysis station and machine number used for each dialysis session and the names of staff members who connect and disconnect the patient to and from a machine.

INFECTION CONTROL TRAINING AND EDUCATION

Training and education is recommended for both staff members and patients (or their family care givers). Training should be appropriate to the cognitive level of the staff member, patient, or family member, and rationales should be provided for appropriate infection control behaviors and techniques to increase compliance. Regulations and recommendations regarding infection control training for health-care workers in general, and dialysis personnel in particular, have been previously published (*180,203--205*). The following recommendations are intended to highlight and augment the earlier recommendations.

- Training and education for all employees at risk for occupational exposure to blood should be provided at least annually, given to new employees before they begin working in the unit, and documented. At a minimum, they should include information on the following topics:

  - proper hand hygiene technique;
  - proper use of protective equipment;
  - modes of transmission for bloodborne viruses, pathogenic bacteria, and other microorganisms as appropriate;
  - infection control practices recommended for hemodialysis units and how they differ from Standard Precautions recommended for other health-care settings;
  - proper handling and delivery of patient medications;
  - rationale for segregating HBsAg-positive patients with a separate room, machine, instruments, supplies, medications, and staff members;
  - proper infection control techniques for initiation, care, and maintenance of access sites;
  - housekeeping to minimize transmission of microorganisms, including proper methods to clean and disinfect equipment and environmental surfaces; and
  - centralized record keeping to monitor and prevent complications, including routine serologic testing results for HBV and HCV, hepatitis B vaccine status, episodes of bacteremia and loss of access caused by infection, and other adverse events. Records of surveillance for water and dialysate quality should also be maintained.
- Training and education of patients (or family members for patients unable to be responsible for their own care) regarding infection control practices should be given on admission to dialysis and at least annually thereafter and should address the following topics:
  - personal hygiene and hand washing technique;
  - patient responsibility for proper care of the access and recognition of signs of infection, which should be reviewed each time the patient has a change in access type; and
  - recommended vaccinations (*206*).

FUTURE DIRECTIONS

Infection control strategies that prevent and control HBV infection among hemodialysis patients are well-established. Areas that need additional research include determining the ideal hepatitis B vaccine dosage regimen for pre- and postdialysis pediatric patients and for predialysis adult patients, as well as the optimal timing for follow-up testing and administration of booster doses among vaccine responders. In addition, further studies are needed to clarify the specific factors responsible for transmission of HCV among hemodialysis patients and to evaluate the effect of the current recommendations on prevention and control of HCV infection in this setting.

Many areas related to bacterial infections in chronic hemodialysis patients need additional information. Studies are needed on the prevalence and epidemiology of bacterial infections among chronic hemodialysis patients and the patient care practices (e.g., those related to vascular access care and puncture) that would be most useful in preventing bacterial infections. Because of the prominent role of dialysis patients in the epidemic of antimicrobial resistance, researchers need to learn more regarding optimal strategies to ensure judicious use of antimicrobials in these patients. Additional topics for future research include determining the frequency of transmission of pathogenic bacteria in the dialysis unit and whether additional precautions are necessary to prevent such transmission.

This document is available on the Internet at <http://www.cdc.gov/hepatitis>. Copies also can be obtained by using the order form at this Internet site or by writing the Hepatitis Branch, Mailstop G37, CDC, Atlanta, GA 30333.

The Evolution of Dialysis Technology

1740s	Peritoneal dialysis first tried on a patient by *Christopher Warrick* in England.
1854	*Thomas Graham*, the Scottish chemist, coined the term dialysis.
1855	*Adolph Fick*, the German physiologist used collodion membranes in diffusion studies.
1877	Peritoneum absorption studied by *G. Wegner* in Germany.
1889	*B.W. Richardson* in England, made what was probably the first reference to the use of collodion membranes in the dialysis of blood.
1913	*John J. Abel, Leonard G. Rowntree* and *B.B. Turner* at John Hopkins University in Baltimore, began to work actively on a dialyzer system and published the technique of "vividiffusion" or hemodialysis. These three men also coined the term artificial kidney.
1914	*C.L. Hess* and *Hugh McGuigan*, Northwester University Medical School, Chicago developed a unique diffusion device that did not require anticoagulants.
1915	*George Haas,* a German investigator, the University Clinic of Giessen, developed multiple layer dialyzers.
1918	*Desider Engel* discovers that protein can pass through the peritoneum.
1920s	*Stephen Rosenak* and *P. Sewon* developed the first catheter.

1923 First clinical application of PD done by *Georg Ganter* in Germany.

1923 *Heinrich Nicheles*, of Germany, used an ox's peritoneal membrane to make an artifical kidney.

1928 *Georg Haas*, incorporated Heparin into his system, becoming the first to use the new anticoagulant for the purpose of dialysis. He is the physician credited with the first use of dialysis on human patients.

1936 Continuous PD done on a patient with urinary obstruction disease until the obstruction was resolved – Wisconsin General Hospital – *J.B. Wear, I. R. Sisk* and *A.J. Trinkle*

1937 *William Thalheimer*, an American scientist, discovered that a membrane used in the sausage industry could be employed in removing solutes from the blood. This man-made cellulose acetate material (cellophane) was uniform in thickness, strong and could be produced in large quantities.

1940 *Willem Kolff,* a young physician working at the University of Groningen in the Netherlands, was one of the first investigators to suggest that toxins might be removed from the blood of patients suffering from renal failure. He developed the first rotating drum kidney to do the dialysis. Sophia Schafstadt became the first patient maintained on dialysis until her kidney function returned (Sept. 11, 1945).

Nils Alwall, in Sweden, developed a vertical-drum kidney, which did not have to be rotated. Later, he introduced the first negative-pressure dialyzer and described the first cannula to be used for gaining access to patient's blood vessels.

1945 First success using PD to treat a patient with acute renal failure – Beth Israel Hospital – *Arnold Seligman, Jacob Fine* and *Howard Frank.*

1946 *M.R. Malinow* and *W. Korzon* did studies on fluid removal from the blood using negative pressure.

E.G.L Bywaters and *A.M. Joekes,* in London modified the Kolff kidney blood-delivery circuit, replacing the original burette with an open gravity-fed version.

1948 *Alfred P. Fishman* and *Irving Kroop* performed the first clinical use of the artificial kidney (Kolff kidney) in the United States at Mount Sinai Hospital, New York City.

Edward Olson, built the Kolff-Bringham kidney.

1949 *Leonard T. Skeggs* developed the first parallel plate dialyzer.

1950 *Arthur MacNeill, John R. Guarino* and *Louis J. Guarino* developed another more advanced form of the Alwall kidney. The most unique aspect of this new system was that the dialyzing fluid path was inside the membrane and the blood on the outside. The Guarino kidney featured a very important design principle – the closed system fluid removal could be accurate determined at any time.

William Y. Inouye and *Joseph Engelberg* developed the "pressure cooker" kidney. Recognizing the functional deficiencies of the rotating and vertical drum designs, they developed a concentrically wrapped coil that incorporated a plastic mesh sleeve to protect the membrane. The pressure-cooker dialyzer was used clinically by *Lewis Bluemle* at the University of Pennsylvania during the late 1950s

1952 *Arthur Grollman* used flexible polyethylene catheter to perform PD. Used gravity flow and suggested 30-minute dwell time each hour – Southwest Medical School in Dallas.

1955 The disposable thin coil dialyzer was developed by *Willem Kolff* in collaboration with Travenol Laboratories (Baxter Healthcare Corporation).

1956 Disposable coil dialyzer unit, arterial and venous blood line, 100 liter tank and sigma motor pump were available for sale from Travenol Laboratories.

1957 *Fredrik Kiil* of Norway, developed the parallel-plate dialyzer. He was interested in producing a dialyzer with a large surface area that required a lower blood priming bolume and could be used without a blood pump.

1959 First PD chronic treatment of patient (7 months) using a permanent indwelling catheter done by *Richard Ruben.*

1960 *Wayne Quinton* and *belding Scribner* implanted their new shunt in the arm of Clyde Shields, one of their chronic dialysis patients. This development enabled Scribner to establish the first dialysis unit for chronic patients in the U.S. at the University of Washington Hospital. This unit had twelve beds. He also developed a central delivery system.

In the early 60s *James Cimino,* Bellevue Hospital, New York City, developed the concept of the internal fistula with the help of *Kenneth Appel.*

1961 *Stanley Shaldon,* in London, successfully trained a patient

to set up his own machine, initiate and terminate dialysis (Self Care Dialysis).
Yuki Nose administered the first home dialysis treatment in Japan.

1961 *Norman Lasker* creates the first Peritoneal Cycler using 2-liter containers and warming of solution prior to patient delivery.

1962 First silicone permanent catheter – *Russell Palmer* and *Wayne Quinton.* Later modified by *HenryTenckhoff.*

1963 In India, *P. Koshy* started hemodialysis at Christian Medical College Hospital, Vellore.

1964 A new blood pump developed by Sarns Company, located in Ann Arbor, Michigan. *Belding Scribner*, trained one of the first patients for home care.

1967 *Richard D. Steward,* developed hollow fiber dialyzer and performed its first clinical trial.

Cuprammonium cellulose tubing became available and was incorporated into a series of dialyzers.

1968 Baxter engineers designed the RSP machines (recirculating, single-pass system) using the experience of *Joseph Holmes* and *Paul Michielsen.*

1970s Manufacturers developed a variety of disposable coil dialyzers.

1973 Federal support for treatment of end stage renal disease became available in the United States.

1974 Large surface-area dialyzers became available which allowed a decrease in the dialysis time.

1975 *Jack Moncried, MD* and *Robert Popovich,* a Biomed Engineer, develop continuus ambulatory peritoneal dialysis (CAPD) at the Austin Diagnostic Clinic in Austin, Texas. The first patient is Peter Pilcher, who is dialyzed for 2 months and then receives a transplanted kidney.

1977 *Karl Nolph, MD* begins dialyzing patients at the University of Missouri using CAPD.

1977 *Dimitrios Oreopoulos, MD* at Toronto Western Hospital uses first 2-liter collapsible bags of peritoneal solution to dialyze a patient. By using a "Y" tubing set, the bad is not disconnected until the drain cycle is completed. By minimzing connections and disconnections, the frequency of peritonitis is reduced expanding the growth of PD.

1979 Baxter Healthcare releases the first complete CAPD system utilizing three different dextrose solutions for control of the ultrafiltration rate, a solution transfer set and a "prep" kit to aid in control of infection.

1980 Automated peritoneal dialysis and continuous ambulatory peritoneal dialysis were widely accepted as maintenance dialysis modalities.

1983 Medicare legislation provides for reimbursement for home dialysis as well as in center therapy. This enhances the expansion of PD into the home.

1989 Recombinant human erythropoietin was commercially available for use in ESRD patients.

2008

Centers for Medicare and Medical Services (CMS) releases conditions to coverage mandating that all existing patient-care dialysis technicians must be certified by April 15, 2010. New Technicians will have 18 months after their date of hire to become certified.

2009

Willem Kolff passes away of February 11th 3 days shy of his 98th birthday. Considered by many to be the "Father of Dialysis", he successfully dialyzed the first patient on September 11th, 1945 until kidney function returned.

Equipment Introduction Dates

COMPANY	MODEL	YEAR	COMPANY	MODEL	YEAR
Travenol	UA10	1956	Cobe	Centry 2000 – Rx	1983
Milton Roy	Model A	1964	B. Braun	HD-secura	1984
Drake Willock	DWS-4002	1964	Travenol	Miro Clav	1984
Drake Willock	DWS-4011/4015	1966	Travenol	SPS-450	1984
Milton Roy	Model B	1967	Fresenius	A1008	1984
Travenol	RSP	1967	Drake Willock	DWS-480	1984
Drake Willock	DWS-4215	1969	Fresenius	2008C	1984
Milton Roy	Model BR	1970	Fresenius	2008D	1986
Marquqrdt	REDY	1972	Cobe	Centrysystem 3	1987
Drake Willock	DWS-4216	1972	Travenol	SPS-550	1988
B. Braun	HD 103	1972	Fresenius	2008E	1988
Cobe	Centry	1972	Baxter	1550	1991
Cordis Corp.	Dialysystem	1975	Althin	System 1000	1991
Cobe	Centry 2	1975	Fresenius	2008H	1992
Extracorporeal	SPS-350	1978	Althin	Ultratouch 1000	1995
Cordis Dow	Seratron	1979	B. Braun	Dialog	1995
B-D / Drake W.	7200	1980	Althin	Tina	1997
Cobe	Centry 2 – Rx	1980	Baxter	Meridian	2000
Cobe	Centry 2000	1981	Baxter	Arena	/2003

The National Association of Nephrology Technicians/Technologist (NANT)
A Historical Perspective

Mark Rolston (edited and condensed by D. Concepcion)

The beginning of NANT grew out of a necessity that was recognized by a small group of fifteen individuals in Philadelphia in 1983. The dedicated individuals realized the importance and contribution of dialysis technicians to the safe, efficient delivery of dialysis. Their vision and goal was the recognition and acceptance of dialysis technicians as an integral and professional member of the interdisciplinary team replete with identity, respect, principles, association and representation in the dialysis community.

Prior to 1983, dialysis technicians were embraced and were members of the American Association of Nephrology Nurses and Technicians (AANNT). In late 1982, a referendum or change in the articles of incorporation occurred in AANNT. The nursing leadership goal was to have nephrology nursing recognized as a formal nursing specialty. The recognition would be from the American Nursing Association (ANA) or the National League of Nursing (NLN). Each of the two organizations had specific rules or regulations that were mandated as prerequisite requirements for attaining recognition. Common to both of the organizations was the rule that all voting members (or members able to influence the association's business), be registered nurses. The rule was not elitist but rather for ethical and/or legal considerations.

The referendum or change placed all non-registered nurse members into the associate member category. Associate member status gives all the benefits of full members except for voting privilege. The inability to vote removed the ability for dialysis technicians to participate in the course, direction, mission, advancement and pursuit of professionalism and identity. As the forefathers recognized that taxation without representation was contrary to the growth of the nation, dialysis technicians/technologist recognized that without a voice there would be no advancement or future for dialysis technicians.

NANT became the voice. NANT began with Stuart Kaufer as President, Karin Nelson as Secretary-Treasurer and a working capital of $500 that was generously contributed by Ron Fuller. NANT's first year was one of uncertainty, lack of organizational structure and depleted funds. Individuals who were supportive of the goal of NANT made personal financial contributions without the promise of reimbursement. NANT survived the first year's trials and tribulations. The first NANT meeting, outside of the national was staged in the Fall of 1983 by Marilyn Urps in Houston, Texas.

C.W. Miller, as president and Karin Nelson, as Secretary-Treasurer, maintained the leadership of the association for the next three years. The articles of incorporation, the association constitution, obtaining the non-profit status

(503C) and the association by-laws were created and obtained. NANT's first chapter was formed in North Carolina in 1985.

The years prior to 1987 were one of apathy and lacked community and technician participation. Interest in NANT was nonexistent to the extreme that at times, only the president, C.W. Miller was the only active participant. Support came in the person of Karen Osband. The two would endeavor to maintain the existence of NANT without members, without active participation and without a working fund.

At the 1987 national meeting in New York City, the future of NANT was in the balance. With the lack of active and consistent participation and interest C.W. Miller announced to the attending few members that at the conclusion of the meeting, NANT would seize to exist. The prospect of NANT dissolving motivated the individuals present to react, take interest and heed the call to service.

The call to service strengthened the leadership of NANT. Significant contributions were provided by Martin V. Hudson, providing leadership as the choice for president, Jeff Hove, from New York as Secretary, and a Texan who possessed a financial knowledge named Larry McGowan. Others who strengthened the board were Philip Varughese from New York, Wayne Bynum and Wes Watkins from North Carolina (who started the North Carolina Chapter), Betty Verbal from Miami, Mike Nelson from Seattle, Jerry Beck and Randy Gates from Arizona and Maurice Kaufman from Chicago. Major contributions were also provided outside of board membership from Anthony Messana, Sally Burrows-Hudson, Doug Leuhman, Doug Vlcheck, Joan Arslanian, Edith Oberly, Jean Kammerer, Anna Corea and Ben Lipps.

Support and contribution to NANT also came from the vendor community. Contributions of expertise and funds were provided by Fresenius, Baxter, Gambro (Cobe), Renal Systems, AMGEN, Althin (Drake-Willock), AUTOMATA and scores of others. Without their support, NANT would have ceased to exist.

With leadership and active participation, NANT became recognized as "the" technician organization and acquired seats on various regulatory organizations (i.e., AAMI, FDA, etc). The task of defining the various technical roles such as, technologist, patient care, equipment technician, and reprocessing began.

A speaker at the 1988 meeting in Reno identified a crucial fact that would challenge dialysis technician and NANT the quest for professionalism, identity and respect. Paraphrasing the speaker, she stated that technicians would never amount to anything due to the lack of goals nor esprit de corps as a profession. In addition, technicians lacked institutes of higher learning, and never received the socialization received by nurses and physicians as to how a professional should act.

The 1989 meeting in Dallas was attended by over 250 members, a small testimonial to disputing the speaker the previous year. But growth had its peculiar problems; there were greater responsibilities, greater expectations, increased work and increased tasks. The existing board lacked adequate numbers to accomplish the work load and the financial stability that at times required personal financial sacrifices. But NANT persevered.

NANT achieved major success in the year (1990-1991) leading up to the San Francisco meeting. The association acquired a seat on the ESRD Coalition (a forum of Renal Care Associations who address issues common to the ESRD Community). The accomplishment was the result of efforts of the sitting President of NANT, Jeff Hover. NANT achieved the seat on the coalition as a result of a motion by the American Association of Kidney Patients (AAKP), seconded by the Renal Physician's Association (RPA) and a positive majority vote by the Coalition member associations. NANT achieved its voice in the community.

Other notable events further advanced NANT. The FDA "Water Treatment Manual" originally distributed by the FDA to the dialysis community was viewed as the "bible" for hemodialysis water treatment. The FDA had exhausted all the available manual and had no plans to print further copies. NANT realized that the unavailability would be a significant loss to the dialysis community. NANT petitioned the FDA for the rights to the document with the intent of reprinting for distribution. The template for the manual was sent to NANT. The original reprinting was published with the financial assistance from Continental Water, Zyzatech and others.

Another achievement was the genesis of NANT's marketing plan. Corporate sponsorship was minimal from a few generous corporate sponsors. Recognizing the contribution of technicians and supporting their advancement was emerging. NANT adopted the philosophy that all technicians deserved any and all education that could be delivered. Opportunities would be available for patient care, reuse or biomedical technicians regardless of membership, affiliation or employer. The underlying value to the marketing plan was that a better educated practitioner would deliver higher quality of care. Dialysis providers, industry, regulatory agencies, and most importantly the patient, embraced the idea.

The 1990-1991 symposium was specially significant in that the keynote speaker was the "Father of Dialysis", Dr. Willem Kolff. Dr. Kolff's inspiring and informative message was possible as a result of Martin Hudson's efforts in convincing Dr. Kolff to participate in the symposium.

Martin Hudson's tenure as president recognized that business management was critical to the success of the association. NANT needed to be run as a business. NANT's emerging success produced new commitments and demands. Successfully meeting the new challenges was critical to the continued existence of NANT. A restructure of the Board of Directors resulted in the elimination of the Vice President-elect positions, the installation of Director of Education (Dr. Maurice Kaufman), Director of Chapter Activities and Membership (Dennis Kennedy), Director of Regulatory Affairs (Mark Rolston), Director of Public Policy (Martin Hudson) and Director of Industrial Relations (Jim Boag). The offices of Secretary and Treasurer were combined into one (Mike Nelson). The western region was divided into the Southwest (Randy Gates) and the Northwest (Tom Suttle). Mark Rolston was voted in as the President-Elect (unopposed).

Marketers, managers, salesperson, technical writers, orators, symposium coordinators and financial resources were critical to business management. The board of NANT recognized that their governing body lacked the skills and resources to maintain the business management of NANT. The business management was given to MOE-TEK*. The firm was an association management company formed by NANT's treasurer, Larry McGowan. With the formal business relationship between NANT and MOE-TEK*, Larry McGowan relinquished his position as treasurer of NANT to become MOE-TEK*'s Executive Director. MOE-TEK would be paid to provide many services sporadically performed by volunteers. Enhanced internal and external communications, financial management with accounting, budgeting and investment, marketing, symposium management, membership date base management, image and industry presence would be provided by MOE-TEK*. Without Larry McGowan and the service provided by MOE-TEK* NANT would not exist.

The existence of NANT can also be credited to contributions and support given by the American Nephrology Nurses Association (ANNA) and its members such as Sally Burrows-Hudson, Barbara Bednar, Gerry Biddle, Ron Brady (ANNA Executive Director) and scores of others. Other significant contributions came from the National Kidney Foundation's Council of Nephrology Nurses and Technicians (CNNT) Jean Kammerer.

In 1991-1992 the ESRD Coalition impaneled the ANNA/NANT/CNNT Task Force on Patient Care Technicians (PCT). The task force was convened to reach consensus on PCT role description, minimum qualifications, educational curriculum and certification. Consensuses on all the items were reached by the task force within three years. Community endorsements on the consensus documents did not materialize and the materials faded into obscurity. Nevertheless, many of these consensus documents actually form the basis for many of NANT's position papers.

During this period a program that would provide a streamlined application process, quick evaluation, tracking and security for CEU's was incorporated. NANT had been certified to grant Continuing Education Units (CEU) by the California Board of Nursing (CBN) in 1989. The program incorporated allowed NANT to retain the rights to accredit not only education symposia, but also dialysis technology schools and industry in-house training program.

For years NANT has been trying to acquire for dialysis technologists/technicians the recognition as a legitimate allied health profession. Originally, the recognition would have come from the Coalition of Allied Health, Education, and Accreditation (CAHEA) an arm of the American Medical Association (AMA) and the Federal Bureau of Health Professions (BHP). Jeff Hover did a tremendous amount of work in this area. The AMA, however, dissolved CAHEA and in its place the Coalition of Allied Health Professions and

Accreditation (COAHPA) was formed. NANT has been in dialogue with COAHPA since its formation. Coincidental to the formation of COAHPA, the federal government contracted with the Pew Foundation to form the Pew Health Commission to study American healthcare and to make change recommendations for the new millennium. Largely through the auspices of Dr. Maurice Kaufman's efforts, NANT has kept abreast of these developments. These developments form an integral portion of NANT's long range strategic goals.

The 1992 Chicago NANT Symposium debut three tracks that incorporated into the program: basic, intermediate and advanced tracks. The symposium delineated the tracks thus allowing attendees the tracks of their choice. The Chicago symposium remains to this day the most successful NANT symposium. Some of the highlights of the meeting included field trips to Baxter's research facility at Round Lake and the City College of Chicago Malcolm X Campus Dialysis Technologist Program (the oldest program of its kind in the nation).

TECHSPO was introduced in 1992 as a means to reach and educate more practitioners by bringing the meeting to the practitioners. The first TECHSPO was in Atlanta and was viewed a tremendous success. The cooperative efforts of NANT and the Renal Physician's Association (RPA), staging the meeting in conjunction with Dialysis Clinic Incorporated's annual chief technician meeting and the manufacturer's of dialysis delivery systems and water treatment vendor's support and a combined meeting management by MOE-TEK* and Nephrology Management Group contributed to the success of the meeting.

Joint symposium management was necessitated by the fact that NANT's business affairs were increasing exponentially. In addition, regional meetings were not part of the MOE-TEK*'s contract. The association was outgrowing MOE-TEK*'s capability. With MOE-TEK*'s request and mutual agreement between the managing firm and the association, a new management firm was sought. The Institute of Association Management Group drafted a "Request for Proposal" for the management of NANT's then 90 member association. Of the 38 proposals received, NANT's board selected Sherwood Groups as the managing form, due primarily for their marketing expertise. By January 1993, NANT had a new association management, a treasury including Certificates of Deposit and there was optimism for growth and stability.

NANT acquired a new professional logo, stationary, professional documents, a presentations booth for shows, an association banner/standard and a professional marketing plan, membership database management with monthly demographic reports and trend analysis that facilitated long-range strategic planning, and a person-to-person call line for assistance. In addition, NANT received a grant from Renal Systems to produce the Core Curriculum for Reprocessing Technicians.

As previously mentioned, NANT had established a dialogue with the AMA's CAHEA and subsequently with COAHPA and the Pew Health Commission in an attempt to acquire status as an allied health profession. Attaining such status had prerequisites. One critical issue was certification. The PEW Commission and the Bureau of Health Professions recommended that the professional allied health associations be the accrediting bodies with either an in-house certification/registration arm or have an independent but closely aligned certification board. The board must be members practicing in the field and the dialysis technician/technologists could not be subject to any other health profession, industry, or other entity's possible or potential influence. The Pew Commission's position was delineating the real and potential ethical/legal conflict-of-interest scenarios that could exist or arise in any other type of relationship.

NANT's long range strategic planning generated many initiatives in order to comply with the Pew recommendations. Ad hoc committees were established to deal with each compliance issue. One of the committees established was to address certification. NANT had no intent to be in the certification business, which would violate their charter as an educational organization. An association of understanding and agreement with the Board of Nephrology Examiners Nursing and Technician (BONENT) certification organization was a plausible strategy. Unfortunately, due to miscommunications, misunderstandings the arrangement between the two organizations did not develop.

The Ad Hoc Certification Committee had continued in its development of articles of incorporation, a constitution, bylaws, position papers and started test item writing activities. In early 1994, NANT's committee ceased to exist and in its place the independent Ad Hoc Committee on Dialysis Technician Certification was formed. This organization evolved into the entity known today as the National Nephrology Technology Certification Board.

In keeping with the philosophy of PAX NANT (NANT Everywhere), the association was involved in 1993 in numerous activities: participated in the revision of the State of California's Dialysis Technician Certification program through the Department of Health, testifying before various state regulatory agencies, commenting on various position papers, and active involvement with the ESRD Coalition, AAMI and the FDA.

The year 1993 also saw some major growing pains which jeopardized the stability of the association. The MOE-TEK* database was not compatible with the Sherwood Group's database. Furthermore, some of the MOE-TEK* database had deteriorated. Expensive time and effort would be required to rebuild the database. In addition, there were other projects that required additional time and money that were originally not budgeted. In 1993, the symposium was held in Orlando, FL. Several factors, including poor attendance, led to the symposium incurring a serious debt for NANT.

Other disastrous events befell the association. Sam Swann, who had won the 1993 president-elect position, resigned in December dues to personal reasons. Attendance at TECHSPO continued to decline, accumulating greater debts. But, the 1993 election provided some new leadership to NANT: Keith Miller (North Central), Larry Byers (Northwest), Joe Sala (Northeast), Melody Devenport-McLaughlin (Southwest), Belinda Bethea (Chapters and Membership Coordinator), and Dan Ghesquiere as Secretary-Treasurer.

Despite the low attendance, the Orland meeting did have a number of significant noteworthy events. The Torchbearer Award debuted and was awarded to fifteen individuals. The award is bestowed upon individuals who either held aloft the spirit of NANT or introduced and encouraged NANT in areas not familiar with NANT. The Lifetime Achievement Award was given to Jeff Hover, Mike Nelson, Sally Burrows-Hudson and Larry McGowan. Also debuting at the meeting was NANT's highest honor, the Martin V. Hudson Award which symbolized a lifetime of significant contributions to the field of Nephrology Technology. The first recipient was rightly bestowed on the namesake of the award, Martin Hudson. The keynote speaker was Colin Aldridge, President of the European Dialysis and Transplant Nurses Association (EDTNA) and European Renal Care Association (ERCA). The NANT Town Hall meeting debuted and was received with a "rousing" success.

In serious debt, NANT's executive committee (President Tom Suttle, President-elect Sam Swann, Immediate-past president Mark Rolston and Secretary-Treasurer Dan Ghesquiere) worked very hard to address NANT's financial situation.

The 1994 election was unique in that every position up for election was contested between multiple candidates. Dennis Todaro (President-elect), Pat Parra (Southwest), David Small (Southeast) would bring new insights and capabilities. The year would bring a change in how NANT would evaluate projects on the basis of conservative revenues; if the project was revenue neutral it was accepted, if there was any potential for loss of money, the project was denied.

The 1994 meeting in Dallas had mixed results. The meeting was financially successful as a result of NANT.. The Dallas budget was extremely conservative, attendance was up and corporate sponsorship funded the many social events. Doug Luehman received the Martin V. Hudson Award. However, other management decisions
resulted in a closed door NANT board meeting without Sherwood Group representatives. The consensus of the board was that the board never felt confidence in Sherwood Group and that Sherwood Group did not have NANT's best interest at heart.
It was decided that in the fall, which ended the second year of the three-year contract, the Sherwood Group would be evaluated per contract. In addition, the Sherwood Group desired a raise in fees. The evaluation meeting with Sherwood occurred in October 1994. The broad view was that neither party was happy with the other. Accordingly, NANT received a formal letter indicating that Sherwood Group would severe its relationship with NANT.

As a contingency plan, Management Excellence Institute (MEI) was contacted in the spring of 1994. MEI's president Fran Rickenbach stated that in whatever capacity NANT desired they would be ready. MEI had been the runner-up to Sherwood during NANT's original search for a management company. Upon receipt of Sherwood Group's letter of termination, MEI was contacted and six days later, the transition had begun. By January of 1995, the transition to MEI management was complete. In the transition, two TECHSPOs were scheduled as planned in February and March, as well as the National Meeting in Philadelphia. Much of the success of the transition was owed to the fact that items, which would take time for MEI to learn, were vested with experienced board members. This allowed MEI to focus on their immediate strengths in data base management, the elections and especially, communications.

The Philadelphia national meeting highlights were Dr. Derrick Latos (RPA president) keynote address that charged technicians to embrace their larger roles and practice the "art of the possible" and Dr. Eli Friedman's tremendous rendition of dialysis past, present and future. Edith Oberly received the Martin V. Hudson Award. The inaugural meeting of NANT's Industrial and Medical Advisory Board occurred, laying the foundation for enhanced communications.

After the Philadelphia meeting, NANT experienced a period of unprecedented growth. Under President Dennis Todaro's leadership NANT solidified itself as the technician organization. Tremendous demands were placed upon the organization such as speaking engagements, business meetings, and position papers, commentaries and especially testimony before various state legislative committees, state nursing boards and chapters. Mr. Todaro continues to represent NANT and act as a resource for state legislative issues such as the Ohio Technician Task Force where he appointed Keith Miller, Mark Parks and James Glenn Johnson to represent NANT and Ohio technicians on the Ohio Legislative Task Force. The task force included participants from the ANNA, Ohio Renal Association (ORC), Ohio Nurses Association, Ohio Renal Physician's Association (ORPA), and representatives from various state legislators including State Senator Grace Drake. In addition, Mr. Todaro steered the TECHSPO seminar back to its original intent: to provide hands-on experience for equipment technicians and real-life practical knowledge for patient care practitioners. Subsequently, TECHSPO was renamed "Advances in Dialysis" to reflect more accurately what the seminar represents. NANT had always been an accrediting body of formal dialysis technician programs. It was under Dennis's leadership that the process was formalized and enhanced.

The 1996 national meeting was held in Anaheim, California. It was a success both in attendance and in content. The highlight of the meeting was the joint ANNA/NANT sessions on team dynamics. Mark Rolston received the Martin V. Hudson Award. Philip Varughese took over as President. This was the first year that NANT offered scholarships to technicians interested in attending the symposium. Over the years, Baxter Healthcare Corporation, AMGEN, Fresenius Medical Care, NA, Integrated Biomedical Technology, Minntech Corporation, RPC, and Serim Research have funded more than 200 technicians' participation at the annual symposium. An anonymous scholarship was set up on 2008 in memory of Tom Blackstone, Dialysis & Technology and a great supporter of technicians.

Philip Varughese has helped NANT achieve recognition both nationally and internationally but his most important contribution has been in membership recruitment. Phil had recognized membership as an important component requisite to NANT attaining its strategic goals. Membership steadily increased first under his leadership as the President of the New York Chapter and secondly, as the National President. In 1997, Philip authored the *Study Guide for*

Dialysis Technologist and the second edition in 1998. Philip Varughese continues to be one of the industry's strongest supporters of certification, education, standardized training, and professional recognition of dialysis technicians.

The national meeting in Minneapolis continued the success of recent years. It included another ANNA/NANT joint session on personal empowerment. Club NANT debuted as a casino-like social event. John Sweeny was presented the Martin V. Hudson Award. Over the years, John Sweeny, Global Training Manager, Baxter Healthcare Corporation has continued to provide NANT with important core education for technicians. He has participated in almost every educational program that NANT has offered to technicians, on both the national and regional level. He has contributed to both NANT publications, *A Study Guide for Dialysis Technologists* and *Dialysis Technology Manual for Dialysis Technicians*, both of which are now considered essential publications for technicians preparing for national certification. John Sweeny's lectures on the dialysis process have taught thousands of technicians over the years and have added greatly to NANT's educational program.

In 1997-1998 NANT continued to achieve success under the leadership of President Jim Curtis. Mr. Curtis was the project manager for *Dialysis Technology* released in Minneapolis. The publication serves as a study guide for clinical technicians. Jim Curtis worked to cement relationships with co-existing educational organizations such as the ANNA, RPA and AAKP. He also was a driving force for greater dialogues with other nephrology organizations.

In 1998, Jim Boag received the Lifetime Achievement Award for his significant contributions to the nephrology technology profession. Shortly after receiving this award, he passed away. In 1999, this award was renamed the Jim Boag Lifetime Achievement Award and was first presented to Jim Curtis. Other recipients include Lorus Hawbecker, educator; Joe Mazzilli, technician; Ron Whisnant, Baxter Healthcare Corporation; Hugh Doss, HDC; Jack Dillon, Medical Solutions International; and Wen Wu, Integrated Biomedical Technology.

There are other members who have made major contributions at the end of the 20th century. Those individuals are Belinda Bethea and Pat Parra in membership and chapter activities. Clifford Glynn, Kathy Myers, Betty Verbal, Joe Sala, Ron Scardino and Bruce Farrow for their participation and representation at regional and national conferences; Larry Byers as Director of TECHSPO and Advances in Dialysis. Finally contribution by carrying the NANT torch to their respective provider, industry and regulatory organizations.

Over the past several years, the aforementioned individuals have provided the energy for NANT's achievements but most certainly, MEI has provided the vehicle by which NANT has achieved success. As a professional association manager, Fran Rickenbach has prevented NANT from making the same mistakes of the past.

In the 21st century, NANT continues to recognize outstanding contributions to the nephrology technology profession. Winners of the President's Award include Susan Hansen, founder of the technician training program and Dr.

Maurice Kaufmann, Malcolm X College; Belinda Bethea, NANT's first woman president; Claudia Douglas, Hackensack Medical University; Lorus Hawbecker, educator; Joe Mazzilli, Hackensack Medical University; Mark Neumann, editor, NNI; Vern Taaffe, RPC; Wayne Carlson, Minntech Corporation and Heather Paradis, Liberty Dialysis.

NANT leadership continues to draw on the strength in the technician community. Presidents included Zelma Griffin, University of Chicago; Clifford Glynn, University of Louisville; Efraim Figueroa, Nipro Medical; Tim Dillon, Kansas City; Danilo Concepcion, St. Joseph Hospital and Forest Rawls, Atlanta.

NANT continues to partner with ANNA and industry leaders on projects. In the aftermath of Hurricane Katrina, NANT served as the distribution point for funds for displaced technicians with the support of AMGEN.

The intent of this historical perspective was to document the contributions made to the ESRD Community by a few dedicated and committed technologists/technicians. Failures were made but, they were far outweighed by the successes. Dialysis technologists/technicians maintain the largest number of professionals in the interdisciplinary delivery of dialysis care to the dialysis patient. A national recognition has been achieved in the April 15, 2008 publication of the Centers for Medicaid/Medicare Services Conditions for Coverage, for a formal requirement for dialysis technician national certification. As a result, NANT has become the focal point for educational preparation material for all of the technician certification exams. Technicians throughout the country are cooperating with dialysis units and ANNA Chapters to organize study courses for certification.

NANT's presence on the web was refreshed in 2008 with the introduction of its new website, developed in collaboration with Technology Management Solutions, Inc.

NANT's presence in the dialysis community is undeniable. NANT continues its collaboration and participation with organizations such as ANNA, RPA, NRAA, AAKP, AKF, CMS, FDA, AAMI, NKF/CNNT, BONENT, NNCC, NNCO and KCP. It has participated in task force and projects such as the FFBI, KCER, and the CMS Community Forum for the conditions for coverage. NANT remains committed to improving the quality of care of the CKD patient as an integral member of the health care team.

National Association of Nephrology Techinicians/ Technologists

PO Box 2307
Dayton, OH 45401-2307

NANT membership *application*

General information *Please type or print*

Date

Name

Home address:

Street address

City / State / Zip

Country

Area code / Phone

Work address:

Position / Title

Employer

Department / Division / Facility

Street address

City / State / Zip

Country

Area code / Phone

Area code / Fax

Preferred mailing address: ☐ **Home** ☐ **Work**

Email

Type of membership you are applying for:

☐ **Full ($50.00)** You must be a staff technician, equipment technician, chief technician or LPN/LVN to be eligible for this voting category of membership.

☐ **Associate ($50.00)** All others, except students, are eligible for this non-voting category of membership.

☐ **Student ($35.00)** Limited to full-time students only; copy of current student ID must accompany application. This is a non-voting membership.

Personal data

What is your gender?

F ☐ Female M ☐ Male

What year were you born?

Are you certified?

1 ☐ BONENT 3 ☐ NNCC
2 ☐ NNCO 4 ☐ Other

How long have you been involved in dialysis? Check only one.

1 ☐ 1 – 5 years 3 ☐ 11 – 15 years
2 ☐ 6 – 10 years 4 ☐ Over 15 years

What best describes your position? Check only one.

A ☐ Staff technician F ☐ RN
B ☐ Equipment technician G ☐ Administrator
C ☐ Chief technician H ☐ Supervisor
D ☐ LPN/LVN I ☐ Student
E ☐ Physician X ☐ Other

What type of organization is your primary employer? Check only one.

A ☐ Hospital/University D ☐ Manufacturer/Supplier
B ☐ Chain affiliation X ☐ Other
C ☐ Free standing unit

In what areas of dialysis are you involved? Check all that apply.

A ☐ Patient care D ☐ Equipment maintenance
B ☐ Reuse E ☐ Transplant
C ☐ Administrative X ☐ Other

In what areas of dialysis are you employed? Check all that apply.

A ☐ Chronic C ☐ Acute
B ☐ Home training X ☐ Other

Method of payment

(Federal ID # for voucher use only: 14-1722307)

☐ Payment enclosed *(make checks payable in US funds to: NANT)*

☐ Invoice against Purchase Order #:

☐ MasterCard ☐ VISA
(this is a ☐ *company card* ☐ *personal card)*

Person's name on card

Card Number Exp. Date

Signature

Sponsor

Chapter

please return this form with your payment to PO Box 2307 • Dayton, OH 45401-2307
phone 937-586-3705 • *toll-free* 877-607-6268 • *fax* 937-586-3699 • *email* nant@nant.meinet.com • *website* www.DialysisTech.NET